Iron

The work on which this publication is based was performed pursuant to Contract No. 68-02-1226 with the Environmental Protection Agency.

Iron

Subcommittee on Iron

Committee on Medical and Biologic Effects of Environmental Pollutants

Division of Medical Sciences
Assembly of Life Sciences
National Research Council

University Park Press
Baltimore

UNIVERSITY PARK PRESS
International Publishers in Science and Medicine
233 East Redwood Street
Baltimore, Maryland 21202

Typeset by American Graphic Arts Corporation.
Manufactured in the United States of America by The Maple Press Company.

Library of Congress Cataloging in Publication Data

National Research Council. Committee on Medical and
    Biologic Effects of Environmental Pollutants.
    Subcommittee on Iron.
    Iron.

    Bibliography: p.
    Includes index.
    1. Iron—Physiological effect.   2. Iron—Environ-
mental aspects.   3. Iron—Toxicology.   4. Iron
metabolism.   I. Title.   [DNLM: 1. Iron.
2. Environment.   QV183 N2775i]
QP535.F4N37   1978   574.1'9214   78-14276
ISBN 0-8391-0126-0

28,453

Contents

Committee on Medical and Biologic Effects of Environmental Pollutants

Subcommittee on Iron

Clement A. Finch University of Washington School of Medicine, Seattle, Washington, *Chairman*

Elmer B. Brown Washington University School of Medicine, St. Louis, Missouri

John C. Brown Agricultural Research Center, Beltsville, Maryland

Howard E. Bumsted U. S. Steel Research Laboratory, Monroeville, Pennsylvania

Merle Bundy U. S. Steel Corporation, Pittsburgh, Pennsylvania

John D. Hem U. S. Geological Survey, Menlo Park, California

J. B. Neilands University of California, Berkeley, California

Darrell R. Van Campen U. S. Plant, Soil, and Nutrition Laboratory, Ithaca, New York

Raymond C. Wanta Bedford, Massachusetts

Munsey S. Wheby University of Virginia School of Medicine, Charlottesville, Virginia

Donald O. Whittemore Kansas State University, Manhattan, Kansas

Thomas H. Bothwell Johannesburg Medical School, University of the Witwatersrand, Johannesburg, South Africa, *Resource Person*

Robert W. Charlton Johannesburg Medical School, University of the Witwatersrand, Johannesburg, South Africa, *Resource Person*

T. D. Boaz, Jr. National Research Council, Washington, D. C., *Staff Officer*

Preface

The Subcommittee on Iron has surveyed the effects of organic and inorganic iron that are relevant to humans and their environment. It must be recognized that the biology and chemistry of iron are complex and only partially understood. Iron participates in oxidation reduction processes that not only affect its geochemical mobility, but also its entrance into biologic systems. Hydrated ferric oxide surfaces have adsorbent properties and may act as reaction sites and catalysts. In biologic systems, the iron atom is incorporated into several protein enzymes that participate in many oxygen and electron transport reactions.

A first consideration is the quantity and form of iron in the earth's crust, in the hydrosphere, and in the atmosphere, and the degree of movement that occurs among them (Chapter 1). Of particular interest is the interaction between inorganic and organic forms of the metal. Iron is brought into the organic cycle through sophisticated mechanisms that microorganisms have developed for converting highly insoluble and unavailable forms of iron into usable ones (Chapter 2). Similarly, the roots of some plants have capabilities for retrieving iron from the soil; matching this affinity to the condition of local soils has led to improvement in agriculture (Chapter 3). Vertebrates in general, despite their high iron requirements for hemoglobin synthesis, appear to be able to achieve satisfactory iron balance (Chapter 4). Humans are the outstanding exception—hundreds of millions of the world's peoples are iron deficient because of inadequate amounts of available iron in the diet (Chapter 5). External factors are believed to be responsible for this borderline balance, and deficiency may thus be considered the major iron-related environmental health problem faced by humans. An account is also presented of a much smaller population that shows iron overload (Chapter 7). The presence of large deposits of ferritin and hemosiderin in parenchymal tissues has been shown to result in damage to several vital organs. Acute iron toxicity has been reported, but only with the ingestion of large amounts of iron salts (Chapter 6). Pulmonary inhalation of iron compounds from industrial exposure has not been shown to be a hazard (Chapters 1 and 8).

Acknowledgments

This document is the result of individual and coordinated efforts by the members of the Subcommittee on Iron. Although each member was responsible for a specific section, as detailed below, each reviewed the work of the others; Chapter 9, the summary, and Chapter 10, the recommendations, represent a consensus of the members of the Subcommittee.

The Preface was written by Dr. Clement A. Finch, Chairman of the Subcommittee. Chapter 1, on iron in the environment, was written by three authors: Mr. John D. Hem wrote the introduction and the sections on iron in the hydrosphere and in solutions and on transport in the hydrosphere; Dr. Donald O. Whittemore wrote the sections on the earth's crust, man-made sources of iron on the earth's surface and in the hydrosphere, and transport in soil and sediments; and Mr. Raymond C. Wanta wrote on iron in the atmosphere and its transport.

Dr. J. B. Neilands contributed Chapter 2, on microorganisms and iron, and Dr. John C. Brown provided the material in Chapter 3, on iron and plants. Chapter 4, on iron metabolism in humans and other mammals, was the joint effort of Dr. Finch and Dr. Darrell R. Van Campen. Dr. Elmer B. Brown was responsible for Chapter 5, on iron deficiency, and Dr. Munsey S. Wheby for Chapter 6, on the acute toxicity of ingested iron. Chapter 7, on chronic iron toxicity, was a collaborative effort on the part of Dr. Thomas H. Bothwell and Dr. Robert W. Charlton; in spite of their location at such a great distance, which made it impractical for them to be full-time members of the Subcommittee, they served very capably as resource persons, and each even managed to attend one meeting of the Subcommittee. These visits were very helpful to the deliberations of the Subcommittee. The short section on carcinogenesis, mutagenesis, and teratogenesis at the end of Chapter 7 was prepared by Dr. Finch.

Chapter 8, on the inhalation of iron, also presents information on the subject of carcinogenesis; this chapter was prepared by Dr. Merle Bundy. In the Appendix, dealing with the analysis of iron in environmental and biologic samples, Mr. Howard E. Bumsted wrote the section on environmental samples and the section on method of analysis; Dr. Van Campen was responsible for the section on biologic samples, and Dr. Finch for the section on human and animal tissues.

Special thanks are due two members of the Subcommittee: Dr. Whittemore, who was the leader of a subcommittee task force appointed by Dr. Finch to deal with problems arising in non-medical portions of the text as the manuscript neared completion; and Dr. Wheby, who led a similar task force to deal with the medically related portions of the text. Appreciation is also due Dr. Orville A. Levander, member of the Committee on Medical and Biologic Effects of Environmental Pollutants, who served as Associate Editor, in which capacity he ensured an adequate review of the proposed final draft by competent anonymous reviewers whom he invited to submit comments before release of the report.

All bibliographic references were checked for accuracy by Ms. Joan Stokes, who also compiled the consolidated reference list. Ms. Avis Berman edited the manuscript.

Free use was made of the resources available at the National Library of Medicine, the National Agricultural Library, the Library of Congress, and the Air Pollution Technical Information Center of the Environmental Protection

Agency. Also acknowledged is the assistance given to the Subcommittee by the National Research Council's Advisory Center on Toxicology, the National Academy of Sciences Library, the Environmental Studies Board, and various units of the National Research Council.

Iron

1

Iron in the Environment

Iron is element number 26 in the periodic table, and has an atomic weight of 55.85. The average isotopic composition is 5.8% of mass number 54, 91.7% of 56, 2.2% of 57, and 0.3% of 58. It is the fourth most abundant element in the earth's crust; only oxygen, silicon, and aluminum are more common. Metallic iron occurs in a few types of rock and is a primary constituent of the earth's core. However, in most rocks and soils it is combined in crystal structures either as divalent ferrous or trivalent ferric ions. The effective ionic radius of the ferrous ion is about 0.80Å. The ferric ion has an effective radius of about 0.67Å. Fundamental aspects of iron chemistry have been described by Nicholls.[595]

Iron occurs in solution in water as Fe(II) or Fe(III), or as inorganic or organic ferrous or ferric complexes. It also can be found in small quantities as a stable colloid or hydrosol, most commonly constituted of small dispersed particles of ferric oxyhydroxide. The terms "ferric oxyhydroxide" and "hydrated ferric oxide" will be used interchangeably to mean a relatively fresh precipitate with a poorly organized crystal structure. Its composition is approximated by the formula $Fe(OH)_3$, and sometimes such materials are referred to as ferric hydroxide, which does not imply specific stoichiometry or crystallinity. Upon aging, this material may achieve the structure of goethite, α-FeOOH, hematite, α-Fe_2O_3, or other polymorphs.

Because of its fundamental importance to human activities, large amounts of iron are extracted from ores each year, and converted to the metallic form. In this form, iron is chemically unstable under earth-surface conditions and is slowly oxidized and converted to ferrous and ferric compounds. The human contribution to the geochemical cycle of iron is important on a global scale. Bowen[94] estimated that the annual amount of iron mined exceeds the amount carried to the ocean by natural rock weathering by a factor of 8.

The chemical behavior of iron in the environment is a function of the following four characteristics of the element:

Iron participates readily in chemical oxidation and reduction processes. These processes influence the geochemical mobility of the element. Iron forms organic and inorganic complexes that affect its solubility in water and its subsequent chemical reactions.

Surfaces of hydrated ferric oxide are active in absorbing other materials and may act as reaction sites and catalysts. Small particles of the hydrated oxide may form colloidal suspensions in water or air and have large surface-to-weight and surface-to-volume ratios.
Iron is essential to biologic processes and is present in all living matter.

The material in this chapter is set forth in terms of properties, sources, and transport processes that influence the environmental occurrence and behavior of iron. Both natural and anthropogenic sources and processes are considered.

NATURAL SOURCES AND THEIR PROPERTIES

The Earth's Crust

Rocks   Iron is common and abundant in igneous, metamorphic, and sedimentary rocks. In most igneous and metamorphic rocks, more iron exists in the ferrous than in the ferric state, whereas the reverse is true for sedimentary rocks. Data for iron concentrations in the earth's lithic (lithosphere) and continental crust have been compiled by Lepp[488] and Parker;[616] selected values from the literature are listed in Table 1-1.

Iron in igneous rocks is found primarily in the ferromagnesian silicates, the most common of which are olivine, $(Mg,Fe)_2SiO_4$, pyroxenes, amphiboles, and iron-containing micas. Important iron-rich accessory minerals are magnetite, Fe_3O_4, ilmenite, $FeTiO_3$, and pyrite, FeS_2. Basic (or mafic) igneous rocks contain more of these minerals and thus more iron, whereas acidic (felsic) rocks contain less (Table 1-1). The average concentration of iron in the lithic crust is close to that of the average igneous rock, because 95% of the crust is composed of igneous rocks or metamorphic rocks derived from them. In the continental section of the crust, the ratio of the amounts of basic to acidic igneous rocks is $1:2$.[616] The thinner oceanic crust consists mainly of basic rocks. Most of the acidic igneous rock outcrops on the continental crust are granitic; most of the basic rocks are basaltic.

The weathering of igneous rocks oxidizes most ferrous iron to the ferric state. The ferric iron is then incorporated into clay minerals or hydrolyzed to ferric oxyhydroxides. During deposition and diagenesis of sediments derived from the weathering products, iron is often reduced and precipitated as sulfides. The average content of iron in sedimentary rocks (including sediments) given in Table 1-1 is close to that of the average continental igneous rock. The main difference between iron from igneous and from sedimentary rocks is the increase in the average ferric-to-ferrous ratio, from 0.5 to 1.2. The average sedimentary rock is based on a calculated ratio of $81:11:8$ for shale:sandstone:limestone;[310] these

Table 1-1.   Average iron contents of the earth's crust and common rocks

Crust or rock type	Percentage by weight		
	Iron (III) oxide	Iron (II) oxide	Total iron
Lithosphere[a]	2.7	5.1	5.8
Continental crust[b]	2.4	4.7	5.4
Igneous rocks[c]			
Ultrabasic	4.6	8.4	9.8
Basic	4.0	7.1	8.3
Intermediate	3.4	4.1	5.6
Acidic	1.6	1.9	2.6
Sedimentary rocks			
Average sedimentary rock[d]	3.5	2.6	4.5
Shale[e]	4.2	3.0	5.3
Sandstone[f]	1.7	1.5	2.4
Limestone[g]			0.38
Average metamorphic rock[h]	1.5	2.9	3.3

[a] Average of values in Poldervaart[634] and Ronov and Yaroshevsky.[672]

[b] Average of values in Pakiser and Robinson[613] and Ronov and Yaroshevsky.[672]

[c] Average of values in Table 93 of *Geochemical Tables*.[673]

[d] From Garrels and Mackenzie.[310]

[e] From Lepp.[488]

[f] From Pettijohn.[629]

[g] From Clarke.[177]

[h] Average of values for crystalline shields in Table III of *Geochemical Tables*.[673]

components comprise about 97% of all sedimentary rocks. Approximate relative abundances of these rocks on continents have been measured as 50% for shale, 26% for sandstone, and 20% for limestone.[630] About 75% of the world's outcrops are composed of sedimentary rocks. Thus they are much more abundant on land surfaces than in igneous rocks. The iron contents of specific types of shales, sandstones, and limestones have been measured.[630, 673]

The elemental compositions of metamorphic rocks are generally similar to the igneous or sedimentary rocks from which they were formed. The average iron content of metamorphic rocks in the earth's crystalline shields is between that of intermediate and acidic igneous rocks.

Soils   Iron is primarily ferric in most soils, although the ferrous state may be predominant in some soils that are flooded and rich in organic matter. The principal iron-containing menerals in soils are the ferric oxyhydroxides, which are set forth in Table 1-2. The most common

Table 1-2.   Naturally occurring ferric oxyhydroxides[a]

Ferric oxyhydroxide	Ideal formula
Amorphous	Indefinite: often represented as $Fe(OH)_3$
Goethite	α-FeOOH
Akaganéite	β-FeOOH
Lepidocrocite	γ-FeOOH
Hematite	α-Fe$_2$O$_3$
Maghemite	γ-Fe$_2$O$_3$

[a] Derived from Langmuir and Whittemore.[465]

are amorphous oxyhydroxide, goethite, and hematite. They occur in soils as small particles, concretions, lateritic crusts, or coatings on clays and other minerals.[208, 602] The first minerals formed from the oxidation of and by hydrolysis of ferrous iron are goethite and lepidocrocite (although less often), because of the relatively low nucleation energy needed. Hematite usually evolves from long-term aging of amorphous material or from dehydration of fine-grained goethite during warm, dry periods.[463]

The presence of organic compounds in soils retards the crystallization of amorphous oxyhydroxides.[698] In organic matter–containing soil horizons, the ferric oxyhydroxide that forms is generally goethite; in environments poor in organic matter, the amorphous material alters faster to hematite. An increase in the organic content in a hematitic soil could cause the dissolution of hematite through reduction and/or complexing of iron by organic compounds. Then the dissolved iron could reprecipitate as poorly crystallized goethite.[698]

Appreciable amounts of ferrous and ferric iron in soils are incorporated in clay minerals as an essential or a minor isomorphous substitute for Mg(II) or Al(III) in the octahedral sites in their structures. Relatively common soil clays containing iron as an essential constituent are nontronite (a smectite), chlorites, glauconite and some illites or hydromicas (clay micas), and certain vermiculites.[155]

Iron can also exist in some soils as unweathered ferromagnesium silicates and in the heavy accessory minerals magnetite and ilmenite. Soils being formed from shales containing pyrite or marcasite, FeS_2, can become quite acidic, because of the oxidation of the sulfides. The low pH allows some ferrous iron to remain in soil moisture before it can oxidize to form ferric oxyhydroxides. Small amounts of amorphous or poorly crystalline strengite, $FePO_4 \cdot 2H_2O$ exist in many acidic soils with high phosphate activity. Under reducing conditions, vivianite, $Fe_3(PO_4)_2 \cdot 8H_2O$, and siderite, $FeCO_3$, form in some soils.[375] Iron salts are generally limited to poorly drained, acid soils where basic ferric sulfates may be present as a yellow mottling or yellow surface crust.[371]

The iron content of most soils ranges from 0.7% to 4.2%.[408] In a series of U.S. Geological Survey studies of 3,026 soil samples in the United States, the total iron content ranged from 0.01% to 13%, with geometric means for the different types of soils ranging from 0.47% to 4.3%.[182] Based on samples collected from a depth of 20 cm for 861 soils, Shacklette *et al.* determined the geometric means and deviations of the iron content for the western United States to be 2.0% and 1.90%, and for the East to be 1.5% and 2.76%.[707] Vinogradov reported iron contents from 0.59% to 10.4% for 16 soils in different zones on the East European plain.[803]

Iron may be relatively uniformly distributed throughout soil horizons, as in the Mollisols, the main types of soils of the prairie and Great Plains grain-growing states. For example, the iron contents of the A, B_1, B_2, and C horizons of a loam in South Dakota were 2.7%, 2.8%, 2.6%, and 2.7%, respectively.[408] Where the parent material is low in iron, or in environments where iron has been leached from the soil, iron contents can be small. Iron contents averaged approximately 0.5% in the A and B horizons of 30 uncultivated garden soils in east central Georgia.[708] In many moist forest soils, especially the Spodosols of the northeastern United States and northern Michigan and Wisconsin, the total iron concentration can reach several percent in the B horizon. Most of it has accumulated as ferric oxyhydroxides and organic complexes, as shown by an average of 3.6% free iron (dithionite extraction) in the B horizons of 10 northern Appalachian Spodosols.[231]

Under prolonged or intense weathering and leaching (generally in hot, wet climates) soils known as Oxisols can form. Iron contents of Oxisols are high, ranging from 15% to 55%, and consist mainly of ferric oxyhydroxides. The higher percentages of iron in soils often referred to as laterites are generally produced during alternating wet and dry conditions that form more stable, less soluble, crystalline oxyhydroxides, and dehydrate goethite to hematite. There are no major areas, however, where recently formed Oxisols exist in the continental United States.

In poorly drained soils high in organic matter, ferric iron can be reduced to the ferrous state. Dissolved ferrous iron can be relatively high in the soil moisture and may be controlled by the solubility of ferrous carbonates, sulfides, or phosphates, depending on the oxidation potential and activities of the anions. Iron contents can vary widely, e.g., from 1% to 20%, as measured in ash of five peats in the United States.[408]

Sediments   Sediments are composed of more ferrous iron than soils because most soils occur above the water table and thus are aerated, whereas many sediments are deposited or undergo diagenesis in anaerobic, subaqueous environments. In a normal depositional environment, in which the pH is generally 8 ± 1 unit, ferric oxyhydroxides are the most important form of precipitated iron. During diagenesis, original amorphous oxyhydroxides and poorly crystallized goethite generally are converted to hematite.[463] Several reviews have discussed the formation of sedimentary iron minerals,[61, 155, 211, 425] and concentrations of iron in average sedimentary rocks are listed in Table 1-1.

Ferrous compounds usually form in sediments in a reducing environment generated by the oxidation of organic matter. Pyrite is the most common one, although siderite can form in waters where relatively high iron, low calcium, and very low sulfide activities exist, as in some nonmarine aqueous environments. During diagenesis of sediments with high organic content, available sulfate, and slow detrital addition, all iron compounds tend to transform into pyrite.[211] Thus, pyrite is common in many clay sediments high in organic matter and equivalent sedimentary rocks, and is often found in coals and associated strata.

Hydrous sheet silicates, glauconite, chamosite, and greenalite also are major constituents of some sediments. They generally form in marine sediments with pore waters of intermediate redox potential, low sulfide and carbonate activity, and available silica and alumina. In general, sediments poorer in iron occur near coasts; iron content of sediments will increase with ocean depth. Berner[61] and James[425] have summarized the expected sequences of iron minerals to be found from fresh water to progressively deeper marine waters. Argillaceous deep-sea sediments tend to be uniform and moderately high in iron content; the average is 5.8% for total iron on a carbonate and water-free basis.[261] Carbonaceous deep-sea sediments[261] and oxide-rich sediments on the East Pacific Rise[207] range up to 11% and 18% iron, respectively. Ferromanganese oxide nodules on the ocean floor have from 1% to 42% iron in them and average 19%, 13%, and 15% iron for the Atlantic, Pacific, and Indian Oceans.[56, 207]

Iron ores   Rock that is economical to mine for iron ranges from 20% to 69% total iron. Ores may be grouped into four classes, according to whether or not the deposits were formed by sedimentary processes, igneous activity, hydrothermal solutions, or surface or near-surface enrichment. The most important ores are the sedimentary iron formations and ironstones, which are distributed throughout the world. The largest and most abundant deposits are around the Atlantic and Indian Oceans.[425, 449, 615]

Sedimentary iron deposits are mainly marine chemical precipitates of iron weathered and transported from land masses. Most of these ore deposits are very old, generally of Precambrian or early Paleozoic age. Some were altered later by metamorphism or enriched by weathering. On the basis of the predominant iron minerals, the sedimentary ores can be divided into four facies groups—oxide, silicate, carbonate, and sulfide.[425]

"Iron formation" is the term for a thinly bedded rock with layers of iron oxides (mainly hematite and magnetite), siderite, and iron silicates (primarily greenalite, a septachlorite) alternating with cherty layers.[425] The average total iron concentration of such rocks is 28%.[488] Iron formations constitute the greatest single source of iron being mined and the largest reserves of iron. The biggest deposits of this type occur in the Lake Superior region of the United States and Canada, the Labrador districts of Canada, and areas in Brazil, Russia, and Australia. The total iron concentration of the ores generally range from 25% to 45%, except where the silica has been leached and replaced by the action of surface and ground waters. These bodies, enriched by secondary processes, are known in the Lake Superior region as "direct-shipping ores," containing soft goethite and hematite and iron contents of 50% to 68%, or the "wash ores" or "semi-taconite."[449] Where the iron formation has been metamorphosed, coarse grains of hematite and magnetite have formed, and the silicates have altered primarily to chlorite, minnesotaite—the iron analog of talc, $Mg_3(Si_4O_{10})(OH)_2$—and stilpnomelane, a sheet silicate with a structure similar to talc.[425] The resulting rock is taconite, the low-grade ore currently mined in great quantities in the Lake Superior region. Small quantities of the amphiboles making up the cummingtonite-grunerite series are found in metamorphosed parts of the iron formations of northern Minnesota (primarily in the Mesabi Range) and northern Michigan.[425] Most of the asbestiform fibers in the tailings dumped by a mining company into Lake Superior and found in the water system of Duluth, Minnesota are comprised of minerals in this series.[597]

The term "ironstone" includes oölitic iron oxyhydroxides (goethite and hematite) and silicates (predominantly chamosite, another septachlorite). Calcite and dolomite are the common constituents of ironstone and its iron content ranges from 20% to 40%. Thicknesses and lateral extents of ironstone deposits are generally an order of magnitude less than those of iron formations. Thus, the reserves have been estimated as at least 100 times less.[449] In certain locales, the near-surface ironstones have been enriched by leaching of calcite. The Clinton Formation, extending from Alabama to New York, is one of the world's most extensive ironstone deposits. However, most of the high grade, secondarily enriched ore has already been mined in the United States. Important deposits are still being mined in northern Europe and coastal Newfoundland.[615]

Other sedimentary iron ores include bog-iron deposits (e.g., accumulations of iron oxyhydroxides in swampy areas and shallow lakes in northern Europe and North America) and blackband and clayband deposits (e.g., the thin layers of siderite found in the coal sequences of Appalachia and Great Britain).[425] Neither of these are of much present or future economic significance.[449]

Iron deposits resulting from igneous activity or formed during replacement by hydrothermal solutions occur primarily as magnetite and hematite. The magmatic segregations of magnetite in Kiurna, Sweden and the hydrothermally enriched iron-formation in the Quadrilatero Ferrifero of Brazil yield high-grade ores (60–70% iron). At present, both deposits are of modest economic importance to the world and contain moderate reserves. The contact metasomatic ores in the Cornwall area of Pennsylvania, which contain about 40% iron, are moderately important sources of iron now, but are potentially a large resource for the United States. Other igneous-related or hydrothermally related iron ores in the United States occur in Nevada, Utah, New York, and Missouri, but are currently of minor economic importance and of small to moderate potential.[449]

The fourth class of iron ores was produced by surface or near-surface enrichment of preexisting low-grade deposits or by the weathering of iron-containing rocks into laterites. The secondarily enriched ores include the important "direct-shipping" and "wash ores" of the Lake Superior region discussed above. Iron-rich laterites generally result from the weathering of rocks high in iron, such as serpentine, under tropical conditions with alternating wet and dry seasons. Such laterites occur in Cuba and the Philippines and commonly contain 40–50% iron, primarily as poorly crystallized goethite. Many laterites contain enough nickel to be mined for that element alone. Although presently only of minor economic importance, the unmined quantities in the world are large. The continental United States, however, has no major lateritic deposits.[449]

Iron in the Hydrosphere and in Solutions

The main features of the aqueous chemistry of iron and the stability of iron compounds that most commonly participate in chemical equilibria can be shown conveniently by Eh-pH diagrams. Eh represents the redox potential in volts, calculated for equilibrium conditions by means of the Nernst equation. Increasing positive values of Eh represent increasing intensity of oxidation. Figure 1-1 is one form of such a diagram, which shows areas of dominance for dissolved ferric and ferrous species, and for two hydroxy complexes of each. The shading of Figure 1-2 shows the areas of stability for ferric and ferrous oxyhydroxides, $Fe(OH)_3$ and $Fe(OH)_2$, siderite, $FeCO_3$, pyrite, FeS_2, pyrrhotite, FeS, and metallic iron, Fe^0, and the field of stability of liquid water at 25 C and 1 atm. Figure 1-2 also illustrates the total solubility of iron as a function of Eh and pH within the field of stability of liquid water, in the system specified for the diagram. Techniques for preparing and interpreting Eh-pH diagrams have been well documented.[309, 359, 362] The system specified for Figures 1-1 and 1-2 includes water, ferric and ferrous ions, sulfur, and carbon, in equilibrium at 25 C and 1 atm, with a total sulfur activity of

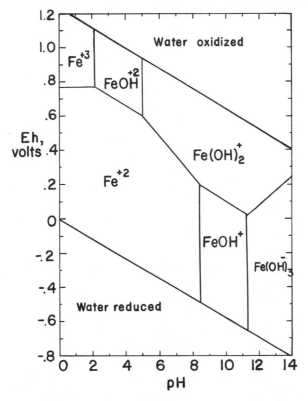

Figure 1-1.   Areas of dominance for dissolved ferric and ferrous ions and hydroxide complexes in dilute solutions at 25 C and 1 atm. Drawing courtesy of J. D. Hem.

$10^{-4.00}$ M and a total activity for carbon dioxide species of $10^{-3.00}$ M. These measurements are equivalent to a sulfate concentration of 9.6 mg/liter and a bicarbonate concentration of 61 mg/liter about the worldwide average concentration of these compounds in river water.[497] The stability fields for solids shown in Figure 1-2 were extended into conditions of low pH in which iron solubility would exceed the highest values shown on the diagram ($10^{-3.00}$ M or 56 mg/liter). For siderite to exist below pH 6, for example, a very high dissolved iron concentration would be required.

Chemical thermodynamic data used in preparing the graphs were selected from Wagman et al.[807] and Langmuir.[464] Their values yield a solubility product of $Fe(OH)_3$ of $10^{-38.49}$. The solubility product for this material can range from about 10^{-37} to about 10^{-44}, depending on the degree of crystallinity.[845] The values chosen represent aged but not dehydrated material. The minimum concentration of iron ($10^{-9.00}$M) shown in Figure 1-2 is 0.56 µg/liter. A relatively low pH or redox potential, attainable in oxygen-depleted water, may stabilize $10^{-3.00}$ M or more of iron in solution. Iron sulfides generally have very low solubilities, but they are unstable in the presence of oxidizing agents. Metallic iron is unstable in water. The species greigite (Fe_3S_4) and mackinawite (FeS) may occur in natural systems, but they are less stable than the forms considered in Figure 1-2.

Silicate rocks commonly contain iron and some iron silicate minerals are important sources of dissolved iron. The iron from these rocks is released during weathering, and subsequent control over the solubility and mobility of iron is a function of processes involving the secondary mineral species shown in Figure 1-2. The silicate minerals may not behave reversibly under earth-surface conditions. The thermodynamic data for iron silicates are incomplete, but some of their effects on iron solubility have been examined.[270]

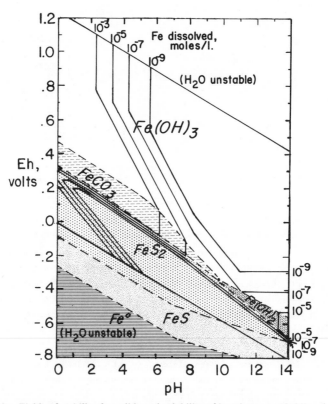

Figure 1-2.   Fields of stability for solids and solubility of iron between $10^{-3.00}$ and $10^{-9.00}$ M in a system of iron, sulfur, carbon dioxide, and water at 25 C and 1 atm. Total dissolved sulfur species activity is fixed at $10^{-4.00}$ M   (9.6 mg/liter as sulfate ion) and total dissolved carbon dioxide species activity at $10^{-3.00}$ M   (61 mg/liter as bicarbonate anion). Drawing courtesy of J. D. Hem.

Complexes of ferric iron with fluoride, chloride, and sulfate are strong enough to affect the chemical behavior of iron in solutions that are enriched with these anions. Organic complexes of iron also may influence iron solubility. In many systems in which organic ligands are abundant, conditions will tend to be reducing, favoring the formation of ferrous rather than ferric complexes. The effects of reduction as well as of complexing must be considered in such systems. Yet some organic ligands are stable in the ferric species regions diagrammed in Figure 1-1. The tartrate ion, for example, forms a strong ferric complex. Ferrous iron–tannic acid complexes and similar substances may be found in swamps and other natural waters that contain organic coloring matter.[763] Some inorganic phosphate compounds of iron have low solubilities and may affect the behavior of the element in such environments as eutrophic lakes, agricultural soils, or sewage. Polynuclear complexes of iron and their relation to biologic iron transport have been reviewed.[728] (The simplified system portrayed in Figures 1-1 and 1-2 considers solely hydroxide complexes.)

As a rule, the redox reactions of ferric hydroxide and ferrous and ferric solutions are fast enough to reach equilibrium readily when the pH approaches neutrality. The oxidation rate of aqueous ferrous iron in aerated water has a second-order dependence on pH,[744] and in acid solutions it is very slow unless some catalytic effect is introduced. Reactions involving sulfide species of iron generally require biologic mediation to attain equilibrium, processes depending on an initial oxidation or reduction of sulfur rather than iron.

When the impositions of an external agent cause the Eh or pH of a solution containing equilibrated concentrations of dissolved iron to change, the system's equilibrium will be upset (see Figure 1-2). Where the necessary reactants are available to restore equilibrium, the resulting chemical processes can serve as sources of energy. Microbiota commonly utilize these sources to catalyze such reactions as the oxidation of ferrous iron to precipitate ferric hydroxide. Where organic material is present, organisms also can mediate reduction of ferric iron or sulfate.

Electron transfers are a part of redox processes, and mechanisms to facilitate transfers are established at active surfaces, such as those of ferric hydroxide precipitates or particles. Such processes can explain coprecipitation of other metals with iron. The sorption of various other dissolved metals by ferric hydroxides is important in controlling concentrations of iron in soil moisture, river water, and underground water.[360] Iron is involved in the synthesis of chlorophyll and its role as an oxygen carrier in electron transfer is essential in the life processes of plants and animals. Therefore, iron is a component of almost all organic matter.

Iron in the Atmosphere

Iron raised from the soil by far exceeds iron residing in the atmosphere from all other natural sources, including meteorites and volcanic eruptions. The Natural Environmental Research Council of the United Kingdom began operating several air sampling stations in sparsely populated areas in the Lake District at Wraymires in April 1970. Most of the samples were analysed by γ-ray spectroscopy. Indications that atmospheric iron at such sites far from industrial activity came from the soil[622] were substantiated in a later extension of the study to other places.[157, 623] Data for about 30 trace and more abundant elements were tabulated. "A remarkable uniformity in the elemental composition of the general aerosol" at seven nonurban sites chosen to represent rural and more industrial exposures was reported.[623]

Iron concentrations ranged from 67 to 940 ng/kg air (about 0.09–1.2 $\mu g/m^3$) during January–December 1972. At Lerwick in the Shetland Islands during June 1972–May 1973, it was only 49 ng/kg air (about 0.06 $\mu g/m^3$). By taking ratios of concentration of iron to scandium in air, and normalizing these ratios by the iron:scandium ratio in average soil (38,000 ppm:7 ppm), an enrichment factor was arrived at to separate elements into those that were "soil-derived," that is, having enrichment factors near unity, and those of "industrial" origin, that is, enriched relative to the soil by a few to many times.[623]

The concentration of iron in surface air layers at nonurban sites in the United States during 1970–1974—represented by the average of intermittent 24-hr measurements throughout the year—ranged from 0.049 $\mu g/m^3$ at a site in Hawaii County, Hawaii in 1972 to 1.091 $\mu g/m^3$ in Park County, Indiana in 1970. The median annual average for 80 complete site-years of record at 38 nonurban sites during the same 5 years was 0.255 $\mu g/m^3$. Seventy-one percent of the annual averages (57 out of 80) fell within the range 0.100–0.399 $\mu g/m^3$. The above statistics are based on analyses of EPA data summaries[795] by the Subcommittee on Iron.

The highest 3-month average of all nonurban site averages during the same period was 2.80 $\mu g/m^3$ at a site in Tom Green County, Texas during the first quarter of 1971. The lowest 3-month average fell below the lowest discrimination levels of the method of measurement (i.e., less than 0.006 to 0.101 $\mu g/m^3$) during one or more 3-month periods at sites in Delaware, Hawaii, New Mexico, South Dakota, and Wyoming.[795]

Atmospheric iron concentrations for earlier years that are based on similar measurements have been summarized.[201] The concentration of iron and its percentage of the total of all species in suspended particulate matter at 10 remote nonurban United States stations for 1966–1967 was 0.15 $\mu g/m^3$ and 0.71%, respectively.[542] The latter figure compares with the lower end of the range of the percentage of iron in soil. In the absence of more direct evidence, it can serve as a rough indicator of the order of the iron fraction (one part in 100 or 200) in soil-derived suspended particulates.

The gross global production rate of tropospheric particulates of all compositions and sizes is estimated by Hidy and Brock[376] to be 10.7 million metric tons/day, of which 94%, or 10 million metric tons/days, is attributed to natural sources. Almost half of the daily production rate from natural sources is attributed to sea spray (3 million metric tons), wind-blown dust (20,000 to 1 million metric tons), forest fires (400,000 metric tons), and volcanoes (10,000 metric tons). The remaining production of tropospheric particulates from natural sources comes from the conversion to particulates of gaseous emissions from vegetation, the nitrogen and sulfur cycles, and volcanoes.[376] Applying the ratio of iron to gross particulate from the paragraph above to the production rate of wind-blown dust, one obtains a rough idea of the rate of iron entry into the atmosphere from this source—100 to 10,000 metric tons/day. Hidy and Brock's estimates[376] of global trospospheric production rates of particulates await corroboration (cf. Whelpdale and Munn[842]).

MAN-MADE SOURCES OF IRON

At the Earth's Surface

Iron ore   Most (95%) crude iron ore mined in the United States and in the world (85%) in 1974 was extracted from open pits because most commercial ore bodies lie close to the surface and have large lateral dimensions. Sites of underground mining in the United States have declined from at least 30 mines in 1951 to six in 1974. The present demand for iron ore of higher iron content and greater uniformity of chemical composition and physical structure requires that almost all crude ores mined be beneficiated before shipment. High-grade ores are crushed and separated according to size fractions. Lower-grade ores are processed more extensively to increase iron concentration as well as to produce the correct physical characteristics for iron extraction processes. After beneficiation, the ores with particles smaller than 0.625 cm are agglomerated by sintering or pelleting. The beneficiated ore, known as usable ore, had an average of 60.6% iron in 1974 in the United States. World production of usable iron ore, iron ore concentrates, and iron ore agglomerates, and their total metal contents for 1974 are set forth in Table 1-3.

Usable iron ore production in the United States has been stable since 1964. Crude ore production, however, has slowly increased because of the rise in the ratio of crude ore mined to usable ore produced. As a

Table 1-3. Worldwide production of iron ore, iron ore
concentrates, and iron ore agglomerates, 1974[a]

Location	Gross weight[b]	Iron content[b]
North America	141,273	86,447
South America	118,062	75,772
Western Europe	135,238	56,387
U.S.S.R.	224,989	132,743
Africa	63,560	38,717
Asia	111,423	59,639
Oceania	99,021	62,518
Total	893,566	511,787

[a] Derived from Klinger.[451]

[b] In thousands of metric tons.

result of mining lower-grade ores, such as taconite, the ratio rose from
1.4:1 in 1954 to 2:1 and 2.6:1 in 1964 and 1974. Of all usuable ore
produced in 1974, 69% came from Minnesota, 13% from Michigan, and
the rest was divided among California, Utah, Wyoming, Missouri,
Pennsylvania, New York, Texas, and Wisconsin (Table 1-4). The ore was
produced by 35 companies operating 66 mines and 44 concentrating
plants.[450] Figure 1-3 shows the locations of principal iron ore sources as
well as major United States ironmaking centers. Approximately one-
third of the usable ore consumed in the United States each year is
imported (Table 1-5). In 1974 Canada supplied 41%, Venezuela, 32%,
Brazil, 14%, Liberia, 6%, Peru, 4%, and other countries, 3% of the
imports. A relatively small amount of usable iron ore was exported,
mainly to Canada and Japan.[451]

The iron and steel industry in the United States is concentrated in
about 20 metropolitan areas. Pennsylvania, Ohio, Indiana, Illinois,
Michigan, and New York produce about three-fourths of the nation's
steel. Iron ores produced in Minnesota and Michigan are usually shipped

Table 1-4. Crude iron ore (<5% manganese) mined in the
United States, by region, 1974[a]

Region	Number of mines	Crude ore[b]	Usable ore[b]
Lake Superior	34	186,886	71,855
Southeast	4	654	292
Northeast	3	7,103	2,396
West	25	26,046	10,664
			500[c]
Total	66	220,689	85,705

[a] Derived from Klinger.[451]

[b] In thousands of metric tons.

[c] Byproduct from processing of other ores.

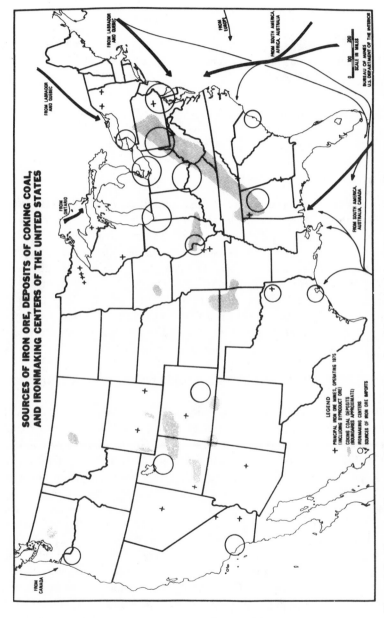

Figure 1-3.   Sources of iron ore, deposits of coking coal, and ironmaking centers of the United States. Reproduced from Klinger.[450]

Table 1-5.   United States iron ore statistics[a]

Usable iron ore (<5% manganese)	Average for 1970–1974[b]	1974[b]
Production	84,934	85,705
Exports	3,194	2,360
Imports for consumption	43,099	48,797
Consumption (iron ore and agglomerates)	134,069	140,371
Stocks[c]		
At mines	13,803	9,555
At consuming plants	51,169	45,971
At United States docks	3,203	3,324

[a] Derived from Klinger.[451]

[b] In thousands of metric tons.

[c] As of December 31, 1974.

by railroad and and lake carriers to consuming centers. However, from December to April, shipment of ores on the Great Lakes is impractical. Thus, ore must be stockpiled, generally in the open, at mines or shipping ports during the winter and at consuming plants during the shipping season to maintain production during the winter. Unprotected piles are possible sources of injections of particles high in iron into the atmosphere. In other states, most ores are transported only by rail. Pipelines are now being successfully employed to move iron ore concentrates as slurries. If pipeline use and winter shipment of ores by lake carriers are increased, steel mills will no longer have to maintain as large inventories of iron ore as before, which will reduce the size of unprotected piles.[450] Amounts of stockpiled ores in the United States are noted in Table 1-5.

Pollution problems related to mining iron ore are mainly associated with large-scale operations. The average ratio tonnage of overburden and waste rock to crude ore mined in the United States in 1974 was 1:1.[450] Most of this waste, some of which is low-grade iron ore, is discharged onto large piles adjacent to the open mine pits. Even larger amounts of wastes result from the beneficiation of the crude ore into usable ore (approximately 130 million metric tons of processing wastes were produced in 1974 in this country). Most of these wastes are piled on the land, although some taconite wastes have been dumped into Lake Superior. Iron oxides, carbonates, or silicates in the waste piles would either remain relatively insoluble or weather to become immobile ferric oxyhydroxides. Only in sulfide-containing rock would there be serious pollution from acidic drainage from mines or ore and waste piles. Other environmental problems are related mainly to the minerals associated with the iron ores, such as the amphibole asbestiform fibers in taconite tailings.[482]

Wastes from bauxite processing   Bauxite, the ore from which alumina and eventually aluminum is extracted, contains from 1% to 25% ferric oxide. Consequently, the residue resulting from alumina production is mainly iron oxide. In the Bayer process, the waste is known as "red mud" and contains about equal amounts of iron oxides and complex sodium-aluminum silicates. The red mud may be fed to a subsequent lime-soda sinter step in the processing of bauxite containing more than 8% silica to produce brown or sinter muds, which are primarily constituted of iron oxides and calcium silicates.[730] About 5.6 million metric tons of these residues are generated annually at eight alumina plants located along the lower Mississippi River, the Gulf Coast, and in central Arkansas.[458] The muds from most of the plants are impounded in large adjacent lakes.[854] Until several years ago, as much as 1.7 million metric tons were discharged annually into the Mississippi River by two alumina plants in Louisiana. Now, one of them has developed a method to eliminate the river disposal. The red mud is first filtered through sand and then stored as reserve landfill. Red mud waste in Arkansas is occasionally used to neutralize acid waste generated at a nearby vanadium plant.[730] In another process, red mud is used to produce a substitute for fluorspar in making steel.[459]

Fertilizers   Iron-containing fertilizer supplements are applied to soils, to foliage in aqueous sprays, or into trunks and limbs of fruit trees. Iron compounds have been applied experimentally in irrigation waters in Kansas.[572] The fertilizer substances used range from inorganic salts to organic compounds and iron chelates, as listed in Table 1-6. Iron salts, e.g., ferrous sulfate, have proved inefficient in enriching many soils because they oxidize and then rapidly hydrolyze to insoluble ferric oxyhydroxides. Ferrous sulfate in solution, however, has proved effective in sprays and is the most popular inorganic salt for foliar application. Fritted iron is not suitable for fertilizing alkaline or calcareous soils, but it can be effective in acidic soils.[777] In addition to ferrous sulfate, the iron compounds used most widely for correcting iron chlorosis are the chelates. Although iron chelates have been applied to soils, they are more effective as foliar sprays. The primary use of chelates is for correcting iron deficiencies in citrus and other fruit trees, grain sorghum, and certain vegetables. Recommended rates of application for aqueous solutions of ferrous sulfate or iron chelates range from 12 to 24 g of iron per citrus tree (as a solution of 0.6–1.0 g iron/liter) to 0.5–1.7 kg iron/ha (as a solution of 6–8 g iron/liter).[572]

In the Hydrosphere

Acid mine drainage   Acidic drainages high in dissolved iron are products of the weathering of iron-containing sulfides in rocks exposed by mining to moist air or oxygenated surface and ground waters. Most

Table 1-6.   Some types of iron fertilizers[a]

Source	Formula	Approximate percentage of iron
Ferrous sulfate	$FeSO_4 \cdot 7H_2O$	19
Ferric sulfate	$Fe_2(SO_4)_3 \cdot 4H_3 \cdot 4H_2O$	23
Ferrous oxide	FeO	77
Ferric oxide	Fe_2O_3	69
Ferrous ammonium phosphate	$Fe(NH_4)PO_4 \cdot H_2O$	29
Ferrous ammonium sulfate	$(NH_4)_2SO_4 \cdot FeSO_4 \cdot 6H_2O$	14
Iron frits	Varies	Varies
Iron ammonium polyphosphate	$Fe(NH_4)HP_2O_7$	22
Iron chelates[b]	NaFeEDTA	5–14
	NaFeHEDTA	5–9
	NaFeEDDHA	6
	NaFeDTPA	10
Iron polyflavonoids	—	9–10
Iron ligninsulfonates	—	5–8
Iron methoxyphenyl-propane	FeMPP	5

[a] From Murphy and Walsh.[572]

[b] EDTA: ethylenediaminetetraacetic acid
HEDTA: hydroxyethylenediaminetetraacetic acid
EDDHA: ethylene diamine di (o-hydroxyphenylacetic acid)
DTPA: diethylenetriaminepentaacetic acid.

acid drainage results from the decomposition of pyrite and marcasite in the coals and associated shales of coal mines. During weathering, the sulfide oxidizes to sulfate, which decreases the pH of the surface moisture or water present. The ferrous iron released forms salts with the sulfate (in drier weather), dissolves in rain or ground waters, or oxidizes. Acidic waters high in iron emerge as seeps from surface mines or as small streams from underground mine portals. The dissolved ferrous iron then oxidizes and hydrolyzes to ferric oxyhydroxides, a reaction that further acidifies the drainage. The oxidation is ordinarily very slow at a low pH, but it is increased by iron-oxidizing bacteria. Dissolved ferrous iron concentrations, which can be higher than 1,000 mg/liter in acidic waters draining from underground coal mines, diminish downstream from the mines as waters become neutralized by dilution or dissolved bicarbonate in non-acidic flows. The precipitated ferric oxyhydroxides coat the stream bed, accumulating as a muddy orange layer that can be as deep as several centimeters near mines.

About 17,600 km of streams have been affected by mine drainage,[531] primarily in the coal-mining regions of Appalachia. Abandoned mines contribute an estimated 60–70% of the drainage, with approximately 85% of that amount from exhausted underground mines.[22] Biesecker and George found that the pH values of many streams containing acid drainage from mines in Appalachia were below 5; 35%, 83%, and 22% of the 318 stream sites sampled had respective iron, manganese, and sulfate concentrations exceeding U.S. Public Health Service drinking water standards.[74] The streams sampled included several sites not affected by mine drainage, in addition to those that were. Future acid mine-drainage problems will be greater in the Appalachian than in the western coalfields, because the former contain coals of generally higher sulfur content.

Prevention methods, such as flooding or sealing mines, have been investigated in attempts to decrease the amount of acid drainage. Many acidic waters are treated by neutralizing them with lime or limestone.[22] One expense of the treatment accrues from the disposal of the voluminous amounts of sludge, consisting primarily of water and gelatinous, amorphous ferric oxyhydroxides.[854] Annual literature reviews on the treatment of coal mine drainage, as well as its chemistry and occurrence, have been published by the Water Pollution Control Federation for over 20 years; a recent one is by Boyer and Gleason.[96]

Steel industry wastes   Iron exists in both dissolved and particulate form in steel industry wastewaters. Water used to clean gases from blast and steelmaking furnaces contains very fine, suspended particles composed largely of iron oxides. Wastewaters from continous-casting and hot-rolling mill operations contain suspended iron oxide scale that is washed from the steel. Surface oxides are also removed in hot rolling and other processes by acid pickling-baths; the spent liquors have high dissolved concentrations.[675, 791, 854] More suspended solids and slightly less dissolved ferrous sulfate, generated in wastes per ton of steel produced, are created with advanced processes for steelmaking than with older technologies. (The newer techniques produce more of certain types of pollutants because larger amounts of water are required per day per ingot ton than for the older processes.[791]) Annual literature reviews of the treatment of wastes from the steel industry have been published by the Water Pollution Control Federation for over 20 years.[722] Wastewater-producing processes, analyses of loads of raw wastes, and water-pollution control practices in the carbon and alloy steel industry have been described.[100]

In the Atmosphere

The iron and steel industry   Man's most important metal is produced and recycled at prodigious and increasing rates. United States and world production of pig iron and steel for 1974, and averaged for the 5-year period 1970–1974, are set forth in Table 1-7.[663] In 1974 the quantity of iron scrap and steel scrap necessary to manufacture pig iron,

Table 1-7.   Iron and steel production in United
States and the world, 1974[a]

	Average for 1970–1974[b]	1974[b]
Pig iron		
United States	83,161	86,615
Worldwide	466,000	514,000
Raw steel		
United States[c]		
Carbon	109,129	114,857
Stainless	1,478	1,950
All other alloys	13,082	15,388
Total	123,688	132,195
Worldwide[d]	641,000	707,000

[a] Derived from Reno.[663]

[b] In thousands of metric tons.

[c] From data of the American Iron and Steel Institute. Includes ingots, continuous-cast steel, and all other cast forms.

[d] Ingots and castings.

steel ingots, steel castings, and foundry and miscellaneous products in the United States was about 10% more than the total pig iron consumed.[239] Several descriptions of iron- and steelmaking technologies and their potential for emission of pollutants are available.[693, 699, 732, 792] In the making of iron and steel, vaporization condensation, fracture, and chemical reaction of raw materials and intermediate products result in the emission of iron and iron-containing particulate matter to the atmosphere, directly or in reduced quantity through pollution control systems. Figure 1-4 shows the principal sources of particulates in iron- and steelmaking processes.

Emissions of iron to the atmosphere from United States iron and steel mills, which can be related to production rates for pig iron and raw steel, are localized, as shown in Figure 1-3 above and in Tables 1-8 and 1-9. Table 1-10 lists raw steel production by furnace type for 1974 and averaged for the 5-year period 1970–1974. Steel production by basic oxygen converter or electric furnace is increasing while production by open-hearth furnace is declining. Production data from cupola furnaces are not available, but in 1974 cupola furnaces consumed 14,190,000 metric tons of scrap and 1,926,000 metric tons of pig iron.[239]

Air pollutant emission factors are ratios of estimates of the average rate at which a pollutant is released as a result of some activity (e.g., industrial production) to an index of the level of that activity (e.g., production rate or capacity). Table 1-11, adapted from data collected for many iron- and steelmaking activities,[793] lists emission factors expressed as weight of particulates per unit weight of metal produced applicable to:

iron and steel mills; ferro-alloy production in electric smelting furnaces; gray-iron foundries; and steel foundries. The efficacy of various pollution control systems in reducing emissions to the atmosphere is indicated in Table 1-11 by reductions in emission factor for a given operation.

That iron does enter the atmosphere from human activities is shown below. A recent survey by Steiner[732] summarizes the relatively few measurements available on the composition and particle-size distribution of emissions (separately) from sinter plants, blast-furnace flues, open-hearth furnaces, basic oxygen furnaces, electric furnaces, and foundry cupola furnaces. Particle sizes smaller than 5 μm may represent half or more of the weight of particulates from open-hearth and electric furnaces, and also from cupola furnaces, especially of the hot blast type. Typical reported[732] iron fractions of particulate matter emitted in ferrous metallurgic processes are summarized in Table 1-12.

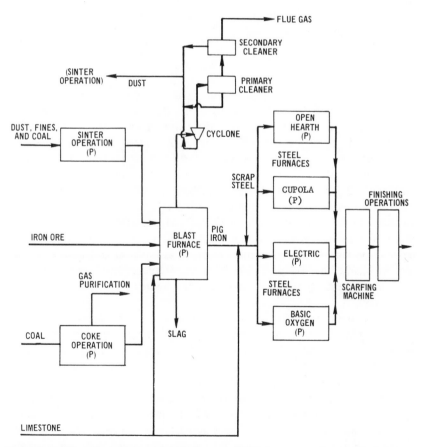

Figure 1-4.   Flow chart of iron and steel processes. "(P)" denotes a major source of particulate emissions if it is not controlled. Reproduced, with revisions, from the U.S. Environmental Protection Agency.[793]

Table 1-8.   United States production of pig iron by state, 1974[a]

State	Production[b]	Number of blast furnaces[c]
Alabama	3,514	9
Illinois	6,517	19
Indiana	15,423	27
Michigan	6,905	9
Minnesota	—	2
New York	4,237	11
Ohio	15,843	41
Pennsylvania	19,681	50
California, Colorado, Utah	4,621	11 (4, 4, 3)
Kentucky, Maryland, Texas, West Virginia	9,872	18 (2, 10, 2, 4)
Total	86,615[e]	197[d]

[a] Derived from Reno.[663]

[b] In thousands of metric tons.

[c] Data from the American Iron and Steel Institute. Total number of blast furnances as of January 1, 1975, 62 of which were out of blast. The total as of January 1, 1974 was 204, 42 of which were out of blast.

[d] Does not include two ferroalloy blast furnances.

[e] Rounded.

Emission factors for carbon monoxide have been estimated for uncontrolled blast furnaces (875 kg/metric ton), sinter discharges (22 kg/metric ton), basic oxygen furnaces (69.5 kg/metric ton), and electric arc furnaces (9 kg/metric ton), as well as for uncontrolled cupola furnaces (72.5 kg/metric ton).[793] Carbon monoxide gas needs to be considered because of the potential for formation of the extremely toxic compound iron pentacarbonyl, $Fe(CO)_5$.[230] Measurements of the presence or absence of such particulates in emissions from iron- and steelmaking seem to be lacking in the literature; cf. treatments of this compound in the petroleum industry and elsewhere.[104, 105] Retention on solid surfaces of low concentrations of iron carbonyl vapors has been discussed recently.[872]

Particulate matter is potentially a carrier for sorbed gases, as is discussed in Chapter 8. It has been estimated that fine magnetite particles can carry one monolayer of sulfur dioxide at 2 ppm sulfur dioxide and 75 monolayers at 66 ppm. Thus some proposed systems of sulfur dioxide control have taken into account the concurrent particulate concentration.

Other sources of iron   Additional sources of iron and its compounds found in the atmosphere as a result of human activities include: mining and handling iron ore; rusting and weathering of exposed iron; weathering of iron pigments in surface coatings; radioactive isotopes of iron; iron-containing fertilizers; decay or burning of vegetation[713] and refuse; and sea spray. None of these sources compares with raising of soil by

Table 1-9.   United States production of raw
steel by state, 1974[a]

State	Production[b]
California	3,895
Illinois	11,738
Indiana	20,945
Kentucky	2,452
Michigan	9,488
New York	4,985
Ohio	22,907
Pennsylvania	30,422
Rhode Island, Connecticut, New Jersey, Delaware, Maryland	6,258
Minnesota, Missouri, Oklahoma, Texas, Nebraska, Iowa	5,217
Virginia, West Virginia, Georgia, Florida, North Carolina, South Carolina, Louisiana	5,097
Arizona, Colorado, Utah, Washington, Oregon, Hawaii	4,465
Alabama, Tennessee, Mississippi, Arkansas	4,325
Total[c]	132,195

[a] Derived from data of the American Iron and
Steel Institute.[23]

[b] In thousands of metric tons.

[c] Rounded.

Table 1-10.   United States steel production by type of furnace[a]

Furnance	Average for 1970–1974[b]	Percentage	1974[b]	Percentage
Basic oxygen converter	66,528	53.8	73,983	56.0
Open hearth	35,162	28.4	32,204	24.4
Electric	21,999	17.8	26,008	19.7
Total[c]	123,688	100	132,195	100

[a] Derived from Reno.[663] Castings produced by foundries not covered in American
Iron and Steel Institute data are excluded.

[b] In thousands of metric tons.

[c] Rounded.

Table 1-11. Emission factors[a]

Source	Total Particulates[c]
Iron and steel mills[b]	
Pig iron production	
Blast furnances	
Ore charge, uncontrolled	55
Agglomerates charge, uncontrolled	20
Total, uncontrolled	75 (65–100, range)
Settling chamber or dry cyclone	30
Plus wet scrubber	7.5
Plus venturi or electrostatic precipitator	0.75
Sintering	
Windbox, uncontrolled	10
Dry cyclone	1.0
Dry cyclone plus electrostatic precipitator	0.5
Dry cyclone plus wet scrubber	0.02
Discharge, uncontrolled	11
Dry cyclone	1.1
Dry cyclone plus electrostatic precipitator	0.055
Steel production	
Basic oyxgen, uncontrolled	25.5 (16–43, range)
Venturi scrubber	0.255
Electrostatic precipitator	0.255
Spray chamber	7.65
Open hearth	
No oxygen lance, uncontrolled	4.15 (2.9–6.0, range)
Venturi scrubber	0.085
Electrostatic precipitator	0.085
Oxygen lance, uncontrolled	8.7 (4.65–11.0, range)
Venturi scrubber	0.085
Electrostatic precipitator	0.175
Electric arc[d]	
No oxygen lance, uncontrolled	4.6 (3.5–5.3, range)
Venturi scrubber	0.09
Electrostatic precipitator	(0.14–0.37, range)
Baghouse	0.045
Oxygen lance, uncontrolled	5.5
Venturi scrubber	0.11
Electrostatic precipitator	(0.165–0.44, range)
Baghouse	0.055
Scarfing, uncontrolled	≤0.5
Electrostatic precipitator	≤0.03
Venturi scrubber	≤0.01
Ferroalloy production[e]	
Electric smelting furnace	
Open furnace	
50% FeSi[f]	100
75% FeSi	157.5

Table 1-11.   (Continued)

Source	Total Particulates[c]
90% FeSi	282.5
Silicon metal	312.5
Silicomanganese	97.5
Semi-covered furnace	
Ferromanganese	22.5
Gray-iron foundaries[g]	
Cupola furnace, uncontrolled	8.5
Wet cap	4
Impingement scrubber	2.5
High-energy scrubber	0.4
Electrostatic precipitator	0.3
Baghouse	0.1
Reverberatory furnace	1
Electric induction furnace	0.75
Steel foundries[h]	
Melting	
Electric arc	6.5 (2–20, range)
Open hearth	5.5 (1–10, range)
Open hearth, oxygen lanced	5 (4–5.5, range)
Electric induction	0.05

[a] Derived from U.S. Environmental Protection Agency data (published in 1973),[793] compiled by R. L. Duprey and R. Gerstle; data for iron and steel mills were revised by W. M. Vatavuk and L. K. Felleisen.

[b] In kg of particles/metric ton of product. Emission factors are based on field measurements of numerous sources.

[c] In kg of particles/metric ton of metal produced, charged, or processed (see notes b, e, f, g).

[d] For carbon-type electric arc furnaces; for alloy-type furnaces, multiply given value by 2.80.

[e] In kg of particles/metric ton of specified product. Emission factors are based on few, if any, field measurements.

[f] Ferrosilicon with 50%, 75%, and 90% silicon, respectively.

[g] In kg of particles/metric ton of metal charged. Emission factors are based on a limited number of field measurements.

[h] In kg of particles/metric ton of metal processed. Emission factors are based on field measurements of numerous sources.

wind, or iron and steel manufacturing, except possibly in highly special circumstances.

For a recent discussion of atmosphere corrosion of iron, including differences between marine and industrial environments, see Spedding.[727] The artificially radioactive forms of iron are Fe-52, 53, 55, 59, 60, and 61. Their radioactivity is induced in naturally occurring iron atoms directly, or in neighboring elements of the periodic table, such as cobalt and manganese, which transmute to radioactive iron by decay. Physical data on these forms of iron are available, and a body of literature exists

Table 1-12.   Iron fraction in industrial particulates[a,b]

Industrial operation	Fe_2O_3	FeO	Fe
Pig iron production			
Blast furnaces			
United States plants			36.5–50.3
European plants			5–40.0
Sintering	11.7–78		50[c]
Steel production			
Basic oxygen			
Noncombusted gas	4.0	21.4	66.7
Combusted gas A	90.0	1.5	
B	80		56.0–57.7
C		11.5–16.4	65.1–68.8
Open hearth	61.3–96.5		55.9–68.0
Electric	19–60	4–11	5–36
Ferrous foundries			
Cupola	d	d	d

[a] Derived from Steiner.[732]

[b] Expressed in percentage by weight.

[c] Typical.

[d] Total of Fe_2O_3, FeO, and Fe components: 5–26% by weight.

on radioactive elements present in the atmosphere during and after testing of atomic and nuclear devices.[257] Superphosphate fertilizers contain 2% iron,[692] some of which becomes airborne during or after application. The iron fraction in true solution in sea water has been estimated at less than 2 μg/liter; the total iron present may be 10 times that quantity.[752] Iron in plankton may account for as much as 16% of the total iron. Thus sea spray associated with surf and wind introduces iron into the atmosphere. There has been some discussion of iron compounds as catalysts in certain chemical reactions in the atmosphere, such as the formation of sulfuric acid mist from sulfur dioxide and sulfurous acid, and the oxidation of nitrous oxide.[727, 837]

TRANSPORT

In Soils and Sediments

Soils The weathering of rocks results in fragmented and/or chemically altered materials that can then form a soil or be transported during erosion as sediment or dissolved solids. Under aerobic conditions, the amount and distribution of iron in most soil profiles are largely dependent on the interactions of the following soil-forming processes:

1. Release of iron from the primary minerals by chemical weathering
2. Movement of iron from the surface to the subsoil by percolation of waters containing iron in solution or colloids

3. Reaction of iron with dissolved and amorphous silica and alumina to form clay minerals
4. Precipitation or accumulation of iron as amorphous ferric oxyhydroxides in the subsoil, followed by slow crystallization to less soluble oxyhydroxides
5. Transport of iron from the subsoil to the surface soil by plant absorption and eventual release through plant decomposition[47]

The rate of the first process generally increases with temperature and moisture. The weathering rate is also dependent on the parent minerals: ferromagnesium silicates and sulfides decompose relatively rapidly, whereas magnetite and ilmenite weather so slowly that they are found as heavy minerals in soils and sediments. Biochemical activity also increases weathering and subsequent release of iron by attack of microorganisms or their metabolites. Slightly acid waters produced by the dissolution of carbon dioxide from organic matter decay also speed up weathering rates. Most iron released from the minerals is ferric, although some organic compounds form chelates with ferrous iron.

The most aggressive and widespread microbial agents are bacterial slimes and acids, phenols, and certain alkaline compounds produced during the decomposition of organic residues or excreted by microorganisms into the soil solution.[26] In addition to mineral (nitric, hydrochloric, and sulfuric) acids, acidic organic compounds are formed. The dissolving action of the latter acid group is sometimes stronger because they often can complex with iron. Organic chelates found in soil solutions are acids of low molecular weight (oxalic, citric, fumaric, and others), fulvic and humic acids of higher molecular weight, and polyphenols.[26, 602]

The weathering of ferrous sulfides in rocks or sediments, e.g., some exposed coals and adjacent shales, is aided by the bacteria *Thiobacillus ferrooxidans*. The oxidation of ferrous iron is a source of energy to these microorganisms. They are most active in an acidic environment (optimal pH range of 1.7–3.5), become inactive above pH 6 and die at higher pH's.[26] Bacterial activity markedly increases the normally slow rate of inorganic ferrous iron oxidation at pH values below 4.

The second process for distribution of iron in soils comes about more rapidly at higher temperatures in highly organic, moist, acidic soils. Although microgram quantities of trivalent iron can move through acidic soils in solution as hydroxy complexes, perhaps most dissolved iron is transported as organic Fe(III) complexes. Divalent iron chelates also are mobilized and transported. Although it has been claimed that some microorganisms reduce ferric compounds,[602] it appears that the organisms isolated are responsible for producing substances (by destroying organic compounds) that then reduce the iron.[26] In addition to the low molecular weight acids, polyphenols, and fulvic and humic acids mentioned above, it is possible that water-soluble hydroxamates formed by soil organisms may complex and transport iron[808] Iron is transported through soil in the form of colloidal ferric oxyhydroxides and iron-containing clays that migrate downward under the peptizing action of dissolved organic compounds.[26, 602]

The third process, iron reacting with silica and alumina to form iron-containing clays, proceeds best at a near neutral pH. Most of the iron reacting is ferric.

For the fourth process, the chemical or biologic environment of many subsoils must cause the iron migrating down into the subsoil to accumulate and stabilize. Heterotrophic bacteria of the Siderocapsaceae family, for example, produce morphologically distinct iron precipitates. The bacteria are widely distributed in soils in which ferrous iron is highly unstable (with near-neutral pH and higher oxygen pressure). These microorganisms utilize or oxidize the organic portion of dissolved iron chelates, releasing the iron that then will precipitate as ferric oxyhydroxides. Any released ferrous iron is rapidly oxidized in such an environment. In very moist or flooded soils with lower oxygen pressure and neutral to slightly acidic pH, bacteria such as *Gallionella* can promote the rate of ferric oxyhydroxide precipitation by increasing the oxidation rate of dissolved ferrous iron existing in ionic or chelated form.[26]

Iron-manganese concretions and iron intercalations are formed by microorganisms at specific soil horizons. In rainy seasons, when soil capillaries are nearly filled with water, iron-accumulating bacteria generally develop in thin zones where soil solutions come into contact with atmospheric oxygen. At moderate humidities, when capillary spaces contain both air and water, iron bacteria develop in dispersed, separate nidi, which become centers of accumulating iron and manganese.[26]

Ferric oxyhydroxides and clays accumulating in the subsoil can also increase the deposition of iron by their absorption of iron-organic complexes. This accretion could continue until the surfaces were saturated and then may be followed by microbial precipitation of iron.[602] The negatively charged surfaces of clays could also adsorb positively charged ferric oxyhydroxide soils that were migrating downward.

Ferric oxyhydroxides deposited inorganically or biogenically in soils are generally amorphous or very poorly crystalline. With aging they can slowly crystallize to more stable, less soluble oxyhydroxides. This has the effect of stabilizing the accumulation of iron in the subsoil.

The rates of the five soil-forming reactions given change primarily with climate and vegetation. In soils (commonly in moist, forested regions) where the first through the fourth processes operate at greater rates than the fifth, iron accumulates in the B horizon. In northern Appalachian Spodosols, the iron accumulated in the upper part of the B horizon is greater under coniferous than deciduous forest.[231] Where chemical weathering is intense and accumulation of iron is more important than its transport, the very high iron contents of Oxisols can result. These soils form in many tropical environments, especially during dry seasons, when the ferric oxyhydroxides are dehydrated to hematite. Oxisols can evolve over a long period of time in a nontropical climate, which has occurred in southeastern Africa.[240] In some moist, acidic soils with a high organic content, the first and second processes may become much more important to iron transport than the third and fourth ones, resulting in small iron concentrations in both upper and lower soil horizons. An oxygen deficiency in these soils slows the decomposition of iron-organic compounds and thus the rate of oxyhydroxide liberation.

When the rate of iron transport is low and the precipitation or stabilization of iron as ferric oxyhydroxides predominates, iron distribution throughout a soil can be relatively uniform. For example, in aerated, calcareous soils in which carbonate equilibria buffer soil pH's near 8, ferric oxyhydroxides severely limit the solubility of iron, and thus its availability to plants. Many dry grasslands soils (Mollisols) and arid soils (Aridisols) are calcareous and cover large areas of the western half of the United States. High phosphate concentrations in a soil can also limit iron mobility. At acidic pH's (where ferric oxyhydroxides are more soluble), ferric phosphates (such as strengite) are more stable than calcium apatites (such as hydroxyapatite) and limit iron solubility.

In poorly drained or flooded soils containing appreciable amounts of organic matter, iron mobility is often much greater than in well-drained, aerated soils. Organic compounds are oxidized, producing an anaerobic environment in which ferric iron is reduced. Concentrations of up to several milligrams per liter of dissolved ferrous iron can exist in the soil solution. The concentrations of other dissolved constituents that could precipitate ferrous iron, particularly carbonate, sulfide, and phosphate, limit the amount of dissolved iron available. Major controls on the transport rate of dissolved iron in flooded soils are concentration gradients for iron and flow of water within the soil.

Soil pH is one of the most important regulators of iron mobility and, by extension, availability to plants. The optimal soil pH for iron supply is about 6.0–6.8 for most plants. Within this range there is usually no iron excess or deficiency.[508] Other important factors, such as the presence of organic chelates or excessive heavy metal micronutrients such as manganese, copper, and zinc, can increase or decrease the availability of iron to plants, respectively.[609]

In addition to controlling the solubility of iron in most soils, ferric oxyhydroxides are important to reactions active at soil surfaces. Specific surface areas of oxyhydroxides are especially large because they can coat particles of all size fractions and thus exert chemical activity much greater than their concentrations would suggest.[429] The isoelectric pH of hydrous ferric oxyhydroxides ranges from approximately 6.7 to 9,[617] indicating a net positive surface for many soils. This surface is shown by the association of anion adsorption in soil clays with the presence of iron and aluminum compounds.[779] Phosphates are fixed by specific adsorption on oxyhydroxide surfaces. Certain trace metals are coprecipitated in or adsorbed on the coatings. Indeed, the hydrous oxides of iron and manganese are thought by Jenne to be the principal control on the fixation of cobalt, nickel, copper, and zinc in soils and freshwater sediments.[429] The adsorption of molybdenum (as MoO_4^{-2}) is stronger on ferric oxyhydroxides than on aluminum oxide and clays, and increases as soil pH decreases for positive oxyhydroxide surfaces.[260]

Sediments   The products of rocks weathering can be transported by the erosional agents water (the most important), wind, and ice. Iron movement in surface running water is principally as suspended matter: colloidal ferric oxyhydroxides, ferric oxyhydroxide coatings on silts and clays, or iron-containing clays or micas. Concentrations of suspended iron thus depend primarily on the total amount of suspended matter and range from <0.1 to $>1,000$ mg/liter. A smaller amount of iron is transported in the bedload of streams in heavy minerals, primarily magnetite and ilmenite. The average concentration of iron in suspended sediments carried by the world's streams to the ocean has been estimated to be 7%.[310] However, when the amount of annual iron flux is compared with the total dissolved and suspended load of material carried by all erosional agents to the oceans, the average iron concentration is 5.5%, about the same as for the average continental crust. Thus, it appears that for the world as a whole, weathering and erosion are not appreciably altering the average iron concentration of the land. Relatively high losses of iron to suspended sediments from land subject to greater erosion are balanced by the accumulation of iron in Oxisols of the tropics.

The concentrations of iron in suspended sediments generally increase inversely with particle size. Coarser sediments deposited near the ocean shore are poorer in iron than those in deeper sea areas, and the highest concentrations are found in red clays far from shore, There are, however, no present-day examples of sedimentary iron deposits approaching in composition and extent the ancient iron formations or ironstones.[425]

Because moisture and temperatures control the movement of iron within aerated sediments, transport tends to be similar to activity found in other soils in the same climate. Where iron mobility is low and the leaching of silica by infiltrating meteoric waters is important, iron formations and ironstones have become enriched in iron, sometimes to depths as great as 400 m in tropical areas.

Most sedimentary deposits are ultimately subaqueous. Iron moves primarily in solution in the pore waters, either by diffusion as a response to chemical gradients or by flow of the water. In lake, estuary, sea basin, or lagoon-bottom sediments, where silts and clays accumulate, often enough organic matter is decomposing to produce reducing conditions. Then ferric oxyhydroxides are dissolved as the iron is reduced to the ferrous state. The ferrous iron can then migrate in solution and be reprecipitated, often as sulfides but also as carbonates and phosphates.

Iron participates in a redox cycle at the sediment-water interface in lakes that become stratified annually. The cycle is an important control on dissolved phosphate concentrations just above the interface. In the summer, phosphate and ferrous iron concentrations increase as adsorbed phosphate is released during the reduction of ferric oxyhydroxides in the oxygen-depleted waters. During turnover of the water in fall, the increase in dissolved oxygen causes ferrous iron to be oxidized. Simultaneously, phosphate concentrations decrease by adsorption on freshly precipitated ferric oxyhydroxides.[839]

Transport in the Hydrosphere

Data on concentrations of iron present in water from various sources are abundant,[60, 497] but they contain certain inherent biases. Common analytic methods are insufficiently sensitive to detect iron concentrations much below 10 μg/liter;[109] hence, extremely low concentrations in fresh water are seldom studied. Concentrations of iron in colloidal particulate form are influenced by the method of pretreatment used in the sampling and analysis procedures.[362, 440] For computing geochemical transport rates, the difference in chemical behavior between the colloidal particulates and dissolved ions is not important, but different forms of iron may behave dissimilarly in plant and animal metabolism. Hence the form in which iron occurs in water requires consideration.

Reported concentrations of iron for the ocean range from a few tenths of a microgram per liter up to about 3 μg/liter.[60] Livingstone [497] reported a mean of 670 μg/liter for "average" river water. Such numbers suggest a solubility higher than that predicted in Figure 1-2. Because oceans and rivers generally contain substantial amounts of dissolved oxygen, the stable forms of iron should be ferric. Filters with pore diameters near 0.45 μm generally are used to remove particulate matter before making analyses, and the dissolved state is therefore functionally defined as any material that can pass through openings of this size. The iron reported in rivers, lakes, and oceans is generally a colloidal suspension of ferric hydroxide particles.[440, 845] In some waters, notably the drainage from swamps or bogs, iron may be present as a stable organic complex. Such waters commonly are yellow or brown and may contain more than one mg of iron per liter. In some such waters, however, the organic matter carrying the iron is also colloidal. Streams contaminated by acid mine drainage have a low enough pH to retain more than 100 mg/liter of dissolved ferric and ferrous species, but chemical reactions with sediment minerals and mixing of the water with alkaline inflows eventually neutralize the acidity as the water moves downstream and the iron is oxidized and precipitated as ferric oxhydroxide.

The range of iron concentrations in groundwaters is quite wide. In systems where organic material is present, the oxygen content of the groundwater can be depleted and iron brought into solution in the ferrous form. Concentrations as high as 50–100 mg/liter can be held in solution in these waters at pH 6.0 and concentrations of 0.5–10 mg/liter are not unusual under many different kinds of geologic conditions.[361] Groundwater with a measurable dissolved oxygen content is likely to contain no more than a few micrograms per liter of iron if the pH is above 5. Pyritic material may be attacked by oxygenated water that recharges the groundwater reservoir, and substantial amounts of divalent iron may be brought into solution as the sulfur of the pyrite oxidizes to sulfate. Once the oxygen is consumed, the ferrous iron can remain in solution.

In 1962, the U.S. Public Health Service set a drinking water standard of a maximum of 0.3 mg/liter of iron, which the 1974 National Academy of Sciences recommendations supported.[580] Most public water supply systems in which iron in the raw water exceeds this limit substantially lower the iron content by treating the water before it is delivered to consumers. However, iron in pipes, tanks, and other equipment in the distribution system exposed to water may be attacked by corrosion. Thus iron occasionally may be present in tap water in greater quantities than allowable. The 0.3 mg/liter limit was set for primarily aesthetic purposes, as water excessively high in iron tends to stain plumbing fixtures and laundry and may be turbid. Compounds like sodium phosphate are sometimes added in water treatment to complex dissolved iron and prevent or delay its precipitation.

Mobilization, transfer, and fixation processes affecting iron in the hydrosphere and controlling its removal from and return to the lithosphere are cyclic. For example, iron released by weathering and biochemical processes from soil and rock minerals is carried to the ocean by rivers and the atmosphere. The average residence time of the iron in ocean water has been estimated to be about 140 years.[317] During this period it participates in chemical and biochemical processes that ultimately precipitate it. The iron next becomes a part of accumulated sediment. The residence time of dissolved iron in the ocean is very brief compared to that estimated for most other metals. Duce et al.[248] reported that the iron content of atmospheric particulates entering the ocean from direct fallout was only slightly enriched above the average level for mineral matter at the earth's surface. They concluded that most iron transported in this way came from natural weathering processes.

The importance of biota in certain redox processes influencing aqueous chemical behavior of iron has already been noted. Some aspects of the topic, however, are worthy of further consideration. Microbes and fungi present in soil and subsurface environments may mobilize iron and bring it into solution by mediating chemical reactions that may serve as energy sources. Pyrite, for example, may be converted to ferrous and sulfate ions; usually oxygen from the atmosphere is the oxidizing agent. The iron is not necessarily oxidized, but it is made available for solution through decomposition of the pyrite. In anoxic environments, ferric oxides or hydroxides may be reduced when certain strains of microorganisms and an organic food source are present.[604]

Once it has been brought into solution, aerobic species of bacteria common in soils and surface waters may catalyze oxidation and precipitation of ferrous iron to ferric hydroxide.[361] This reaction is thermodynamically favorable in the presence of oxygen and thus can be an energy source. The ferric hydroxide produced by the aid of "iron bacteria" such as *Crenothrix* or *Leptothrix* is commonly encrusted on the cells' organic sheaths. Colonies of these microorganisms constitute gelatinous masses of such material and deposits are made in many wells, pipelines, or other places in which water that contains ferrous iron is exposed to oxygen.[603] The deposits are a nuisance in agricultural tile drains in some parts of Florida.[290] These iron-oxidizing bacteria have some characteristics of fungi.[603]

Growth cycles of marine and fresh water algae and related aquatic microorganisms substantially influence concentrations of iron in open water bodies. Demand for iron by algae during an algal bloom can greatly reduce concentrations in solution. This iron is released again upon death and decay of the plants.[604] Some systems may exist in which increased iron availability can trigger an increase in algal growth. Possible correlations have been advanced between "red tides" noted along the west coast of Florida and flooding in coastal streams that was responsible for carrying organically complexed iron into the seawater offshore (see also Chapter 3).[443, 530]

Transport in the Atmosphere

Direct transport from source to receptor   The atmospheric dispersal of contaminant gases and particles of negligible settling speed is often treated as simply the eddy diffusion of matter from point or line sources.[132, 294, 721, 743, 783] The diffusion model in predominant use applies the continuity equation in a steady (mean) wind and permits the emergent plume to spread crosswind vertically and horizontally while traveling downwind, so as to yield at any given distance Gaussian profiles of concentrations of contaminant in those two directions. Standard deviations of the two crosswind distributions of contaminant that have been widely accepted were derived indirectly from a number of field experiments; they increase with distance from the source and vary with atmospheric stability and with the duration of time over which the concentrations are averaged. Values for these standard deviations ("diffusion parameters") are chosen to obtain estimates of contaminant concentration, despite the absence in most places of direct measurements of atmospheric stability, by schemes that employ one or more of the meteorologic elements commonly measured or otherwise known or estimated—wind speed, solar irradiation, and cloud cover, together with the time of day and sometimes the season.[540] Another approach to these standard deviations of contaminant distributions makes use of the measured range of fluctuations in wind direction at a suitable height above the earth's surface when these data are available.

The empirical basis of the diffusion model for estimating contaminant concentrations at distances beyond a few tens of kilometers is weak, and other methods must be considered.[294, 721, 842] Other problems of the model include its inapplicability in near calms or during precipitation, and the all-too-common absence of a reliable climatologic lexicon of diffusion parameters for the height, terrain geometry, and surface cover, as well as for the geographic location being considered.[818] Some success has been realized in accounting for jet and buoyancy effects on the downwind profiles of plume height versus distance, an understanding important because of the major influence on ground-level contaminant concentrations that the effective height of emission exerts, especially in the near field, i.e., within 20 or so chimney heights.[106, 561,743] Another important effect on plume height arises from the distortion of airflow by buildings and other structures in the vicinity of the place of emission, in an extreme case even resulting in recirculation to the origin. Approaches to estimating the effect of architecture and construction on the performance of relatively short stacks and chimneys have been proposed.[132, 333, 507, 788] Meanwhile, allowance for or even recognition of this effect, e.g., in the fitting of pollution models to measurements, is spotty.

As particles increase above 10–39 μm in aerodynamic diameter, deposition on the earth's surface and thus loss from the diffusing plume must be taken into account. This phenomenon is commonly treated by regarding the particulate plume to be similar to a gaseous plume that is tilted downward at an angle whose tangent is the ratio of the settling speed of the particle to the wind speed,[743] the angle of tilt changing with particle size.

Thus the estimation of contaminant concentrations resulting from specified emission rates of, for example, iron-containing particles, is common practice, but one whose results should be accepted with reserve. The consequences of plausible alternatives to the model inputs advanced from meteorologic and other perspectives might be useful to contemplate, and if even fragmentary measurements of contaminant concentrations are available, patterns of correspondence and dissimilarity should be noted for the information they provide on the dispersal process.

Reentrainment of deposited dust and raising of soil by the wind Contaminants not only enter the atmosphere from stationary and mobile sources of emission, but enter or reenter when winds are strong enough to raise the soil, sand, or previously deposited dust.[133, 153] Little air movement is required when the particles are fine and dry. The presence in the eastern United States of dust from the Great Plains in periods of drought and the sighting of Sahara dust far out to sea are not rare. As a rule, as the wind speed increases, the concentration of airborne dust near the surface will rise—a result of increasing flux of particles from the surface—but then it will fall as dilatation begins to dominate production. With huge reservoirs of dry dust available, the high dust concentration and low visibilities characteristic of dust and sandstorms are reached. This mobility of iron-containing soil and deposited dust is necessarily reflected in the measurements of airborne iron in remote areas discussed above. On local scales, particulate loss from ore piles may be controlled by use of moisture, covers, and wind breaks. The process of reentrainment has been studied and described for an important case by Sehmel and Lloyd.[703]

Observed ambient concentrations Measurements of iron concentrations in surface air from the National Air Surveillance Network between 1970 and 1974 were summarized earlier for nonurban sites. Similar data for 210 urban sites comprising 604 complete site-years of record show annual average values ranging from 0.23 $\mu g/m^3$ at a site in St. Petersburg, Florida in 1972 to 12.12 $\mu g/m^3$ at a site in Steubenville, Ohio in 1971. The median annual average was 1.295 $\mu g/m^3$ or about five times the median annual average for iron at nonurban sites. (The above statistics are based on analyses of EPA data summaries[795] by the Subcommittee on Iron.)

Figure 1-5 shows the cumulative frequency distribution of the annual average iron concentrations over the 604 site-years of record at urban sites from 1970 to 1974, and also corresponding data for the 80 site-years of record at nonurban sites previously noted. The highest 18 (3%) concentrations for urban sites came from 14 places in Ohio, Indiana, Pennsylvania, and Kentucky (cf. Figure 1-3). The lowest 18 came from 13 sites all located in coastal states, including Hawaii, with the single exception of one site-year at Lincoln, Nebraska.[795]

The highest 3-month average was 16.0 $\mu g/m^3$ at the Steubenville site in the second quarter of 1971; the second highest, 14.6 $\mu g/m^3$, was measured at a site in East Chicago, Indiana during the same period. Both sites are near major iron and steel mills. The tendency observed in the data toward maximum concentrations of iron in air during the first half

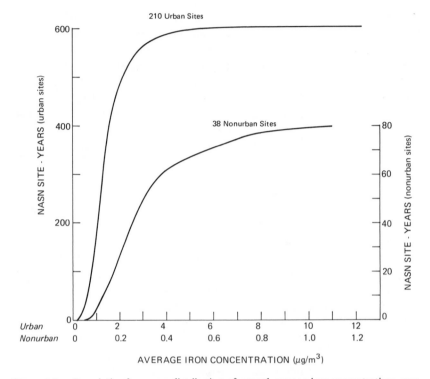

Figure 1-5. Cumulative frequency distribution of annual average iron concentrations over 604 site-years at urban sites and corresponding data over 80 site-years at nonurban sites, 1970–1974. Derived from data of the U.S. Environmental Protection Agency.[795]

of the year is consistent with the expectation that strong winds affect the contents of surface air.

The primary 24-hr and annual mean standards for *total* suspended particulate matter are 260 and 75 $\mu g/m^3$, respectively. Based on the EPA's composite annual average data from 1,096 sites monitored between 1970 and 1974, ambient concentrations of total suspended particulates declined from 80 to 66 $\mu g/m^3$ during the 5 years studied.[797] The percentage of air monitors reporting values exceeding the primary standard decreased from 16% to 8% for the 24-hr average, and from 50% to 23% for the annual average.

Of 236 air-quality control regions (one or more monitors) reporting minimum acceptable data in 1974, 99 exceeded the 24-hr primary standard and 111 exceeded the annual primary standard. "Nonpoint sources"—defined as sources emitting fewer than 90 metric tons/year of total suspended particulates—also contributed to the excessively high amounts.

It is interesting to note that attainment and maintenance of these standards are adversely affected by two kinds of fugitive emissions: those generated from industrial operations and released to the atmosphere through plant apertures other than the primary exhaust system; and those generated by the force of the wind or human activity on the land, including wind-raised particulates from croplands, unpaved roads, and exposed areas at construction sites, as well as surface material made airborne by vehicles and machines from streets,[703] croplands, etc. Fugitive dust is a major problem in the arid West, but it is not confined there. Nationwide estimates of pollutant emissions have declined steadily in tonnage of particulates between 1970 and 1974.[797] One should expect the "soil-derived" fraction of iron in the ambient air to increase or decrease in quantity independently of the decline in emissions, according to cultivation practices and the frequency of dry, windy conditions.

2

Microorganisms and Iron

Iron is thought to be a universal requirement for microbial cells, whether they be prokaryotes or eukaryotes, i.e., bacterial or fungal species.[586] The only possible exception is the lactic acid bacteria. These organisms flourish in environments, such as dairy products, that are notoriously low in absolute and/or available iron concentrations, and they do not contain any cytochrome or hydroperoxidase enzymes.[154] The absence of catalase affords a quick presumptive test for colonies of lactic acid bacteria growing on agar surfaces. In lactic acid bacteria, the normal iron-containing ribotide reductase is replaced by a functionally equivalent vitamin B_{12} enzyme system. Hence, in these species, iron is not even required for synthesis of DNA. Although it is a generally nutritious element for microbes, certain hazardous ecologic and environmental factors are associated with microorganisms and iron and they need to be evaluated.

METABOLISM

Assimilation

All known aerobic and facultative anaerobic organisms possess multiple systems for acquiring this crucial nutrient. These systems fall into two general classes: low affinity, or nonspecific, and high affinity, or specific.[590] In the first system, almost any inorganic or organic iron compound will suffice to support growth, provided it is furnished in substantial amounts. The biochemical mechanism of operation of this pathway is not well understood, because no convenient way exists to probe the biochemical genetics of the system. In contrast, the high-affinity system depends on the elaboration of specific carrier molecules, called siderophores (formerly, siderochromes), and the cognate surface receptors for the iron-laden form of the carrier. Formation of siderophores and their specific receptors may be induced by growth of the organism in environments low in either absolute or available iron concentrations.

Physiologically Active Iron Compounds in Microbes

As a rule, microorganisms contain most of the iron compounds found in the cells of higher plants and animals. Ferredoxins, nonheme iron-containing compounds that serve as electron transfer agents, are abundant in all species requiring a carrier of low-potential electrons, such as the *Clostridia*, and in all organisms performing photosynthesis or nitrogen fixation.[505] It is probably physiologically counterproductive for microbes to synthesize oxygen carriers or possibly even iron storage compounds. However, hemoglobin-like pigments and ferritin have been found in certain species.[223] As expected, siderophores have not been detected in strict anaerobes or lactic acid bacteria. Table 2-1 lists iron compounds reported to exist in microbial species.

Dissimilation

Biologic degradation of microbial iron compounds should be similar to that of higher species, athough the subject does not seem to have been investigated systematically. An organism, *Pseudomonas* FC-1, has been isolated and shown to be capable of growing on ferrichrome A, a fungal siderophore that was its sole source of carbon and nitrogen.[823] Intracellular microbial iron compounds usually are biologic catalysts and are present in small amounts within the organism. No special problems with pollution or other ecologic disturbances are anticipated in the biodegradation of these substances and no toxic substances should be produced.

DEFICIENCY

General Effects of Iron Limitation

Among the aerobic and facultative aerobic microorganisms, the general effect of iron deprivation will be the switch of the metabolism to a more anaerobic character. Glycolysis proceeds without utilizing iron catalysts; the latter begin to influence energy metabolism at the level of the tricarboxylic acid cycle (aconitase, succinic dehydrogenase) and the respiratory chain (heme and nonheme iron). Thus, a facultative anaerobe like *Escherichia coli* has a substantially higher iron requirement when offered succinate rather than glucose as an energy source. Microorganisms seem to be infinitely adaptable to the hardships imposed by living in media low in iron.[551] Certain species of *Clostridia*, when grown without iron, produce a low molecular weight flavoprotein (flavodoxin) with the same redox potential and biologic function as ferredoxin. The metabolism possible at low levels of iron has been characterized.[492]

The production of numerous commercial pharmaceutical products depends on microbes grown at low levels of iron. Formation of citric acid by *Aspergillus niger* and riboflavin by *Eremothecium ashbyii* are examples of commercial fermentations based on growth under iron-poor conditions.[551] Iron represses the synthesis of a number of bacterial

Table 2-1.   Iron proteins and enzymes in microorganisms[a]

Substance	Source
Oxygen-binding proteins	
Hemoglobin-like proteins	Restricted distribution
Leghemoglobin	*Rhizobia*
Iron storage	
Ferritin	*Phycomyces blakesleeanus*
Phosphoproteins	Unknown
Hydroperoxidases	
Catalase	Widely distributed in aerobes
Peroxidase	Restricted distribution
Superoxide dismutase[b]	Widely distributed in aerobes
Electron-transfer proteins	
Heme proteins	
Cytochromes	Widely distributed
Nonheme proteins	
Iron-sulfur proteins	Widely distributed
Nitrogenase	Nitrogen-fixing species
Glutamate synthase	*Escherichia coli*
Hydrogenase	Restricted distribution
Ribotide reductase B_2	*Escherichia coli*
Iron flavoproteins	
Dihydroorotate dehydrogenase	*Zymobacterium oroticium*
Succinic dehydrogenase	Widely distributed
Nitrate reductase	*Micrococcus denitrificans*
Xanthine oxidase	*Micrococcus lactilyticus*
NADH[c] dehydrogenase	*Azotobacter vinelandii*
Malate vitamin K reductase	*Mycobacterium phlei*
Adenylylsulfate reductase	*Desulfovibrio vulgaris*
NADPH[d] sulfite reductase	Enteric bacteria
Formate dehydrogenase	*Pseudomonas oxalaticus*
Oxygenases	
Heme type	
Tryptophan dioxygenase	Widely distributed
Nonheme type	
Diverse substrates	Widely distributed
Iron-activated enzymes	
Aconitase	Coincident with tricarboxylic acid cycle
D-Altronic acid dehydrase	*Escherichia coli*
L(+)-Tartrate dehydrase	*Pseudomonas* sp.
Unclassified iron enzymes	
Enzymes acting on amines, amino acids, and other substrates	Widely distributed
Phosphodiesterase?	Coincident with cyclic AMP

[a] From Neilands[586] unless otherwise noted.

[b] Zinc, copper, and manganese, as well as iron, have been detected in superoxide dismutases from different sources.[334]

[c] Nicotinamide adenine dinucleotide, reduced.

[d] Nicotinamide adenine dinucleotide phosphate, reduced.

toxins.[10-12] The mechanism for inhibiting synthesis of the *Corynebacterium diphtheriae* toxin has been investigated.[571]

Siderophores

Chemical nature, mechanism of action, and distribution among species Siderophores are the low molecular weight (about 500–1,000), high-affinity carriers found in practically all aerobic and facultative anaerobic microorganisms in which they have been sought.[241, 438, 586-591] These compounds are virtually ferric-specific, have large formation constants for trivalent iron (about 10^{30} or higher), and the liganding atoms are generally all oxygens.[587] The complexes are in d^5 orbitals and exhibit rapid exchange kinetics. Chemically the siderophore ligands are usually classified as hydroxamates or catechols. The former are common to higher microorganisms, such as fungi, whereas the latter are usually encountered in the prokaryotes. Ferrichrome, a cyclic hexapeptide ferric trihydroxamate, is the prototype of the hydroxamate class of siderophores. Ferrichrome and ferrichrome A, and the related compounds ferrichrysin, ferricrocin, ferrirubin, and ferrirhodin, are growth-promoting iron chelates. Hydroxamate siderophores are found in *Neurospora*, *Ustilago*, and other basidiomycetes, and in species of the mold genera *Aspergillus* and *Penicillium*. Enterobactin, also called enterochelin, is the prime example of the catechol class and it is produced by the enteric bacteria. The latter, however, may also produce a hydroxamate, such as aerobactin.

Except for the mycobactins, siderophores tend to be exceptionally water-soluble. They can usually be extracted into either benzyl alcohol or phenolchloroform (1:1, vol/vol) from water or aqueous salt solution at neutral or acidic pH. Diluting the organic phase with diethylether and back extraction with water returns the siderophore to the water. If a siderophore does not have a hexadentate coordination (e.g., rhodotorulic acid), it may form polynuclear complexes with iron that are practically impossible to extract into an organic solvent. When this happens, it is best to isolate the substance in the absence of iron. Similar considerations apply to crystallization of the siderophore or its metal complex. In general, the ligands and chelates may be crystallized from the lower alcohols. The hydroxamates are much more stable than the catechols, but ferric hydroxamates decompose upon excessive dry or steam heat. Mineral acid hydrolysis of the ferric hydroxamate will lead to disproportionation of the hydroxylamino group and extensive degradation. A variety of techniques are available for removing iron: extraction with 8-hydroxyquinoline, treatment with dilute alkali in the cold, or continuous extraction with ether in acidic media. The hydroxamic acid group is weakly acidic and has a pK_a of around 9. The hydroxylamino function is easily oxidized at neutral or alkaline pH; it has a pK_a close to

5. The paramagnetic ferric ion causes line broadening in the nmr spectra, but the resonances of the aluminum or gallium complexes are sharp and well resolved.[498, 499]

The hydroxamate siderophores give the Czáky test for bound hydroxylamines; respond (inexplicably) to the Folin-Ciocalteu phenol reagent;[747] are oxidized by periodate to the acyl moiety and a *cis*-nitrosoalkane dimer, and by performic acid to two carboxylic acid fragments; are reduced (if not too hindered) by Raney nickel and hydrogen gas at 22.5 kg pressure; are hydrolytically reduced by hydriodic acid; and are cleaved by nonreducing mineral acid to the carboxylic acid and hydroxylamine components. The hydroxylamine moiety will reduce alkaline tetrazolium without heat, thus affording a sensitive spray for chromatograms.

The catechol class of siderophores give the Arnow test, and those containing the dihydroxybenzoyl nucleus have a characteristic blue fluorescence and an electronic absorbancy band near 310 nm in the ultraviolet region.

The hydroxamate siderophores are well-detected by chromatography on silica gel, either in columns or on thin layer plates.[390] Enterobactin, however, is more difficult to discern by chromatography; 5% ammonium formate (wt/vol) in 0.5% formic acid (vol/vol) will move it slightly on cellulose thin layers or paper, and in 6% acetic acid it has an Rf of 0.29. The iron complex is retarded chromatographically on cellulose weak anion exchangers or on Sephadex LH-20.†

In both catechol and hydroxamate siderophores, the ferric ion is attached to an asymmetric, bidentate five-membered ring. The metal-binding center may be chiral, provided there is optic activity in the ligand; geometric isomers are also formed.[658] Ferrichrome A, ferrichrysin, and ferric mycobactin have been examined by X-ray diffraction and shown to have the Λ-*cis* configuration, whereas ferric enterobactin is probably Δ-*cis*. Isostructural Cr(III) analogs are employed for assignment of configuration and in biologic transport experiments for elucidation of the preferred optic isomer.

The electronic absorption spectra of the ferric hydroxamate siderophores display hypsochromic and bathochromic shifts that are dependent on pH.[587] The red-orange to tea-colored 3:1 complexes have a maximum absorbancy at 425–440 nm with molar extinctions of approximately 3,000. Upon acidification, *tris*-ferric acethydroxamate will decompose through the 2:1 to the 1:1 complex. The 1:1 complex is purple, with a maximum absorbancy near 500 nm and an extinction of

---

† Specific products and trade names are cited solely for illustrative purposes. Mention does not imply an endorsement from the National Academy of Sciences or the National Research Council.

about 1,000. In contrast, a trihydroxamate, such as the ligand of ferrichrome, will retain the $3:1$ structure even at pH 2.

Ferric enterobactin is a wine-colored, water-soluble, trivalent anion with a molar absorptivity of 5,600 at 495 nm. The ferric ion stabilizes the ligand, although the molecule can suffer oxidation of the phenolic hydroxyls and hydrolysis of the ester bonds in the inner ring.

Both enterobactin and deferriferrichrome undergo a drastic conformation change upon metal complexation,[500] which enables the membrane receptors to recognize only the metal-laden form.

Although a very large number of molecular species transport iron in ferrous-ferric microbial systems, predictably, the *tris*-catechol and *tris*-hydroxamate ligands are the most efficient carriers of the ferric ion. A microbial product qualifies as a siderophore if:

It has a high formation constant (about 10^{30} or greater) for Fe(III), high relative avidity for Fe(III) over Fe(II), and kinetic lability.

Biosynthesis can be induced when iron in the microorganism's environment is low.

It is capable of transporting iron in microbes naturally or artificially lacking high-affinity iron transport systems.[40]

Substances with classic siderophoric chemical structures can be assigned to one of seven families. Of these families, one is constituted of catechol and the other six are hydroxamic acids. The figures and captions in this chapter supply formulas and diagrams of the prototypes and a few other members of each of the seven families. A more detailed list of structures has been published elsewhere.[588]

The prototype of the catechol family is enterobactin, the cyclic trimer of 2,3-dihydroxy-*N*-benzoyl-L-serine, shown in Figure 2-1. Ferric enterobactin has been isolated from all enteric bacteria examined for its presence. A compound of related structure, providing two catechols and one salicylic acid moiety bound to spermidine, has been isolated from *Micrococcus denitrificans*.[757]

The ferrichromes, set forth in Figure 2-2, are cyclic hexapeptides comprised of a tripeptide of ferric δ-*N*-acyl-δ-*N*-hydroxyl-L-ornithine and a tripeptide of small, neutral amino acids such as glycine, serine, or alanine. To provide the "hairpin turn" and the cross-β structure in the cyclohexapeptide, one residue must always be glycine (R^1 = H in Figure 2-2). The acyl group furnishing the R in Figure 2-2 may be acetate or some small carbon piece derived therefrom, such as *trans*-β-methylglutaconic acid, as in ferrichrome A, or anhydromevalonic acid, as in ferrirubin or ferrirhodin. In ferrichrome itself, R is methyl and $R^1 = R^2 = R^3 = $ H. Albomycin (grisein) has a close structural relation to ferrichrome; it also displays broad-spectrum antibiotic activity.

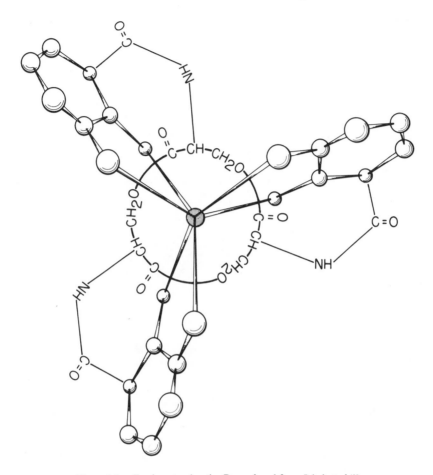

Figure 2-1.   Ferric enterobactin. Reproduced from Isied *et al.*[405]

The rhodotorulic acid family have in common the diketopiperazine of N-acyl-δ-N-hydroxyl-L-ornithine. Members of the rhodotorulic acid family—rhodotorulic acid, dimerumic acid, and ferric coprogen—are diagrammed in Figures 2-3, 2-4, and 2-5.

The citrate-hydroxamate family (Figure 2-6) of siderophores has at least six members, three of which have been isolated. The remaining three theoretical members are awaiting discovery, hence the trivial name "awaitin."

The mycobactin family (Figure 2-7), produced by mycobacteria, has many individual members because of the different substitutions in the R groups, as marked in Figure 2-7. The molecule has six potentially optically active carbon atoms, noted as a, b, c, d, e, and f in Figure 2-7.

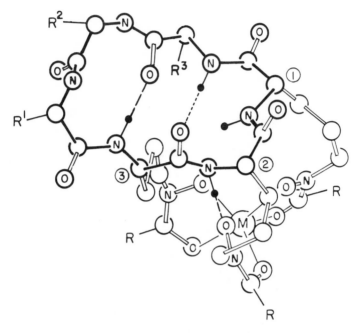

Figure 2-2.   Basic structure of the ferrichrome cyclohexapeptide ferric trihydroxamates. In ferrichrome itself, which has a triglycine sequence, R^1 = R^2 = R^3 = H and R = CH_3. Reproduced from Neilands.[588]

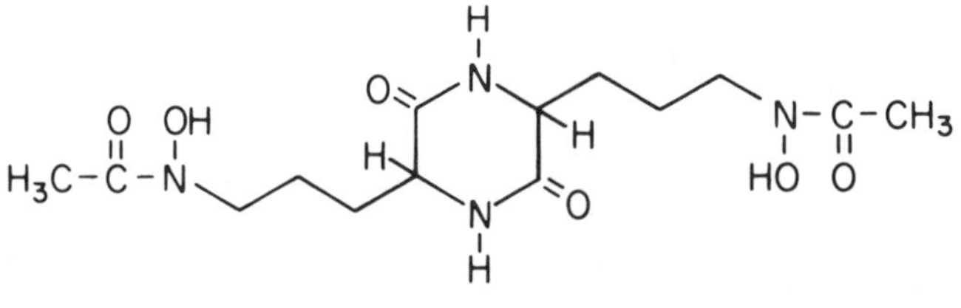

Figure 2-3.   Rhodotorulic acid. Reproduced from Neilands.[588]

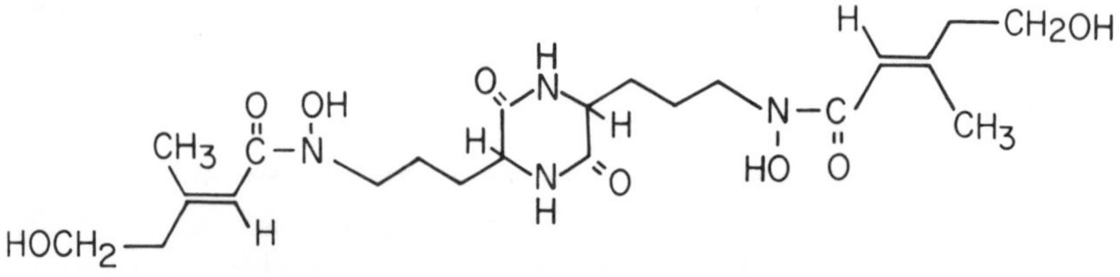

Figure 2-4.   Dimerumic acid. Reproduced from Neilands.[588]

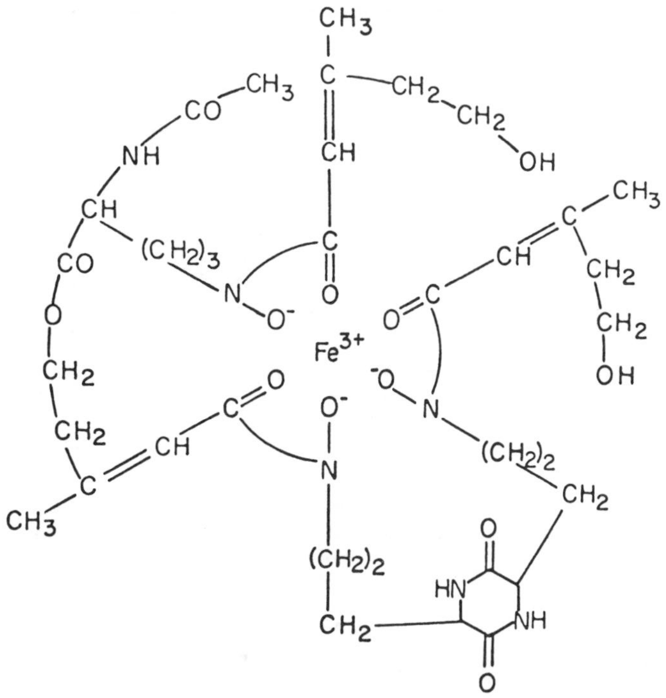

Figure 2-5.   Ferric coprogen. Reproduced from Neilands.[588]

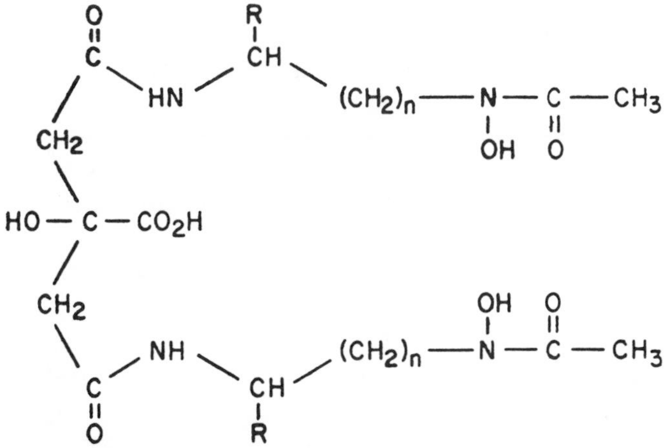

Figure 2-6.   The basic structure of the citrate-hydroxamate family. For the three known members, schizokinen, aerobactin, and arthrobactin, R = H, or COOH and n = 2 or 4. Similar structural perturbations at R and n would afford three additional hypothetical members: awaitins A, B, and C. Reproduced from Neilands.[588]

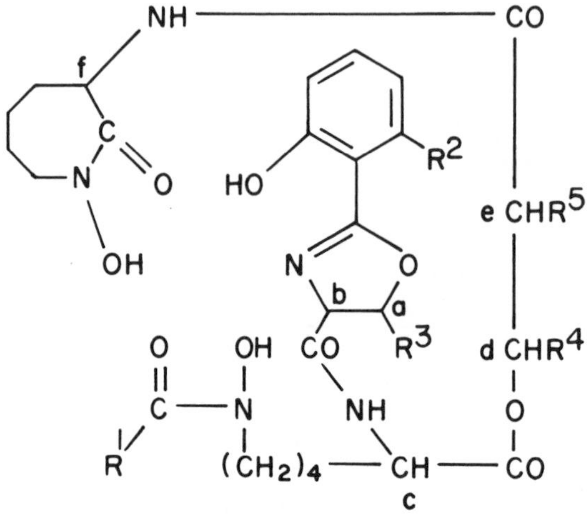

Figure 2-7. The basic structure of the mycobactin family. Depending on the source organism, different substituents occur at R^1–R^5. The ligand has six centers for potential optical activity (a-f). Reproduced from Neilands.[588]

The structure of the fusarinine family is shown in Figure 2-8. Fusarinines (also called fusigens) are obtained from species of fusaria and other fungi.

The ferrioxamines, represented in Figure 2-9, are linear or cyclic molecules containing repeating units of 1-amino-ω-N-hydroxyaminoalkane and succinic or acetic acids. The antibiotic ferrimycin is a further elaboration of ferrioxamine B.

Citrate can be considered as a primitive siderophore in that it tends to form polynuclear complexes at ligand/ferric ion ratios $<20:1$.[728]

The most noteworthy feature of the siderophores is that iron regulates their biosynthesis.[307] No matter what the level of available iron is, it is likely that there can be a constitutive level of siderophore synthesis. However, at less than about 1 μg-atom iron/liter, the substances are leaked to the medium and, in the case of rhodotorulic acid, the siderophore of the red yeast *Rhodotorula pilimanae*, production reaches almost 10 g/liter.[28] In addition to the siderophore itself, the appropriately

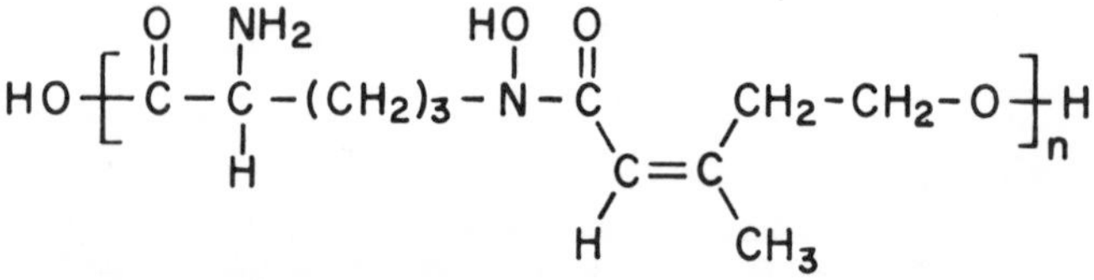

Figure 2-8. The fusarinine family. In the structure shown, n may mean a 1-, 2- or 3-linear arrangement, or a 3-cyclic arrangement. A corresponding series is acetylated on the amino group. Reproduced from Neilands.[588]

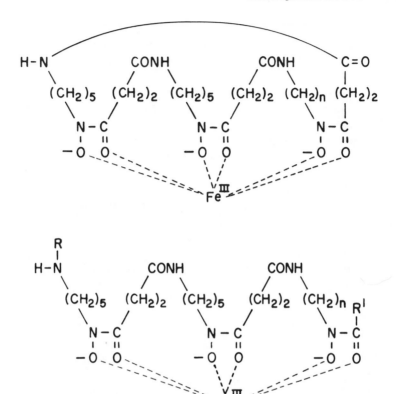

Figure 2-9. Cyclic and linear forms of the ferrioxamine family. In the linear structure, various substituents occur at R and R[1]; n may be 4 or 5. Reproduced from Neilands.[588]

specific membrane receptor proteins are also overproduced by the culture when environmental iron concentrations are low. Such activity may not be true for the ferrichrome receptor in enteric bacteria; these organisms synthesize only enterobactin and not ferrichrome.[541]

Biochemical genetic experiments with *Salmonella typhimurium* prove conclusively that the biologic functions of the siderophores are to solubilize and transport iron.[635] Thus, mutants that are unable to biosynthesize enterobactin can only be grown in citrate medium after addition of excess iron or an organic agent that can decompose ferric citrate, a complex which *Salmonella* cannot transport. Several mechanisms exist to deliver siderophore iron in enteric bacteria.[590] It may be donated to the cell surface or, alternatively, the intact chelate may penetrate the organism. In both instances the iron is liberated by reduction, regardless of whether the ligand is hydrolyzed. The requirement for reduction may obviate transport of nonnutritious metal ions such as Al(III). The presence of the ferric complex in the cytosol would provide the cell with a specific small molecule that could act as an effector to signal the biologic compatibility of the external environment. Various ferrichromes and coprogen can act as germination factors in *Neurospora crassa*.[390]

Working with *Ustilago sphaerogena*, Emery[265] proved that ferrichrome could shuttle iron into the microorganism. The Cr(III) complex is also incorporated by this organism, showing that the Λ-*cis* configuration is one of the actively transported species.[487]

Table 2-2 summarizes the known distribution of siderophores among species; the list is growing rapidly.

Phage-bacteriocin interaction Siderophores have been shown to protect sensitive cells of enteric bacteria against certain phages and bacteriocins ("protein antibiotics" or "killer proteins").[827] Thus ferrichrome, although it is not made by *Escherichia coli*, protects that organism from phages T_1, T_5, and $\phi80$, from colicin M, and from albomycin, a structurally related antibiotic, by a mechanism known to be one of adsorption competition for a common outer membrane receptor.[509] Enterobactin protects the organism against colicin B by the same

Table 2-2.   Known sources of siderophores[a]

Compound	Sources
Phenolates	
Enterobactin (enterochelin); Cyclo-tri-2,3-dihydroxy-*N*-benzoyl-L-serine	*Aerobacter aerogenes* *Escherichia coli* *Salmonella typhimurium*
2-*N*,6-*N*-di-(2,3-dihydroxybenzoyl)-L-lysine	*Aztobacter vinelandii*
2,3-Dihydroxy-*N*-benzoyl-L-serine	*Aerobacter aerogenes* *Escherichia coli* *Salmonella typhimurium*
2,3-Dihydroxybenzoylglycine (itoic acid)	*Bacillus subtilis*
2,3-Dihydroxybenzoylthreonine	*Klebsiella oxytoca*
N^1*N*^{8}-*bis*-(2,3-dihydroxybenzoyl) spermidine	*Micrococcus denitrificans*
2-Hydroxybenzoyl-*N*-L-threonyl-*N*^{4}-[*N*^{1}*N*^{8}-*bis*-(2,3-dihydroxybenzoyl)] spermidine	*Micrococcus denitrificans*
Tri-2,3-dihydroxybenzoyl serine, methyl ester	*Escherichia coli*
Hydroxamates	
Ferrichrome family	
Ferrichrome	*Ascomycetes* *Basidiomycetes* Fungi imperfecti including *Aspergillus, Penicillium,* and probably *Verticillium*
Ferrichrome A	*Ustilago maydis* and *sphaerogena*

Table 2-2.   (Continued)

Compound	Sources
Ferrichrome C	*Cryptococcus melibiosum*
Ferrichrysin	*Aspergillus melleus, terreus,* and *oryzae*
Ferricrocin	*Aspergillus fumigatus, humicola, nidulans,* and *versicolor*
Ferrirubin	*Paecilomyces varioti, Penicillium variable,* and *Spicaria*
Ferrirhodin	*Aspergillus nidulans* and *versicolor*
Albomycins	*Actinomyces griseus*
Form δ_2	*Streptomyces subtropicus*
Form ϵ	
Form δ_1	
Grisein	*Streptomyces griseus*
Ferribactin	*Pseudomonas fluorescens*
Sake colorant A	*Aspergillus oryzae*
Verticillins	*Verticillium dahliae*
Rhodotorulic acid family	
Rhodotorulic acid	*Leucosporidium scottii* (allotype, mating type a)
	Rhodosporidium toruloides (mating type)
	Rhodotorula glutinis
	var. *dairenensis* (type)
	var. *glutinis graminis* (type)
	rubra (type)
	rubra (type)[b]
	Sporidiobolus
	johnsonii (type)
	ruinenii (type)
	Sporobolomyces
	albo-rubescens (type), *hispanicus, pararoseus* (type), and *roseus*
Dimerumic acid (dimerum acid)	*Fusarium dimerum*
Coprogen (compound XFe)	*Neurospora crassa, Penicillium*
Deacetyl coprogen	*camemberti, chrysogenum, notatum,* and *urticae*
Citrate-hydroxamate family	
Schizokinen	*Bacillus megaterium*
	Anabaena (blue-green alga) sp.
Terregens factor (arthrobactin)	*Arthrobacter pascens*
Aerobactin	*Aerobacter aerogenes* 62-1
Mycobactins	*Mycobacteria*
Fusarinines (fusigens)	*Fusarium roseum;* other species of *Fusaria, Aspergillus, Gibberella* and *Penicillium*
Ferrioxamines	
Linear ferrioxamines:	
Ferrioxamine B	*Micromonospora*
	Nocardia

Table 2-2.   (Continued)

Compound	Source
Ferrioxamine	*Streptomyces pilosus* and other *Streptomyces* species
Ferrioxamine D₁	
Ferrioxamine G	
Ferrioxamine A₁	
Ferrimycin A₁	
Cyclic ferrioxamines	
Ferrioxamine E (ferric nocardamine)	*Nocardia* *Streptomyces pilosus* *Streptosporangium roseum* *Chainia* *Chromobacterium violaceum*
Ferrioxamine D₂	
Miscellaneous ferrioxamines	
Metabolite C	
Ferrioxamine A₂	
Other hydroxamates	
Aspergillic acids	*Aspergillus flavus*
Aspergillic, neoaspergillic, muta-aspergillic, hydroxyaspergillic, and neohydroxyaspergillic acids	*Aspergillus sclerotiorum*
Mycelianamide	*Penicillium griseofulvum*
Pulcherriminic acid (2,5-diisobutyl-3,6-dihydroxy-pyrazine-1,4-dioxide)	*Candida pulcherrima* and related yeasts *Fabospora ashbyi, dobzhanskii,* and *lactis*
Hadacidin	*Penicillium aurantio-violaceum* F4070b and *frequentans*
Actinonin	*Streptomyces* sp.
2,4-Dihydroxy-7-methoxyl-1,4-bezoxazin-3-one (DIMBOA)	Certain tissues of higher plants, such as corn seedlings and seedlings of higher grasses; lettuce; leaves of tomato, cauliflower, leek and cabbage; and extracts of carrot roots. Mycelium of *Streptomyces* sp.
Viridomycin A	*Streptomyces viridaris*
Thioformin (fluopsin)	*Pseudomonas fluorescens*
Ferrous-ion-binding compounds	
Pyrimine	*Pseudomonas* GH
Ferroverdin	*Streptomyces* sp.

[a] Compiled from Diekmann,[241] Murphy *et al.*,[573] and Neilands.[586,587,589,590] The organisms listed are typical sources of specific siderophores; however, they may be synthesized by other species. Thus enterobactin is generally produced by the enteric bacteria.

[b] Originally known as *Rhodotorula mucilaginosa*.

mechanism of competition for the enterobactin receptor.[826] In *Salmonella typhimurium*, phage ES 18 and ferrichrome similarly compete for a surface receptor. These protections are quite specific because the various agents are binding to a similar or closely related locus in the receptor complex.

Siderophores, synthetic ferric chelates, and inorganic iron compounds generally protect *Escherichia coli* against colicins B, V, and Ia.[826] This process is not totally understood, except that it is not one of adsorption competition for the outer membrane receptor. The mechanism may amount to repression of the surface receptor by iron, or the metal may somehow interfere with how the colicin works. However, siderophores provide extremely potent protection, which is likely to be of biologic importance.

Guinea pigs kept on a synthetic diet high in iron fare poorly. The gastrointestinal tract becomes distended and it represents a large fraction of the carcass weight (G. Briggs, personal communication). The addition of as little as 5 mg/kg enterobactin to the diet enhances weight gain[445] and restores the gastrointestinal tract to its normal fraction of the carcass weight. It is highly likely that the enterobactin alters the balance of the microbial flora of the gut or the metabolism of the organisms therein. Enterobactin has been shown to display pacifarin activity (nutritional immunity) in mice challenged with virulent strains of *Salmonella typhimurium*.[825]

The siderophore-phage-colicin interaction may determine aspects of how sewage systems work, but basic research on this problem has yet to be done.

Transport of actinide elements   It is estimated that 90 metric tons of plutonium exist in the world, manufactured in the past quarter century by fission reactors. Siderophores represent one of the few natural ligand systems that could coordinate plutonium effectively and initiate its insertion into the food chain.[138] Plants have been shown to concentrate plutonium from the soil. Animals kept on a low-iron diet have been found to take up plutonium more efficiently; once in the bloodstream, the metal is carried by transferrin. It may be excreted as the citrate chelate. It is a possiblity that this man-made element can be moved by natural ligand transport systems intended for iron.[138]

Distribution in nature   Assays with *Salmonella typhimurium enb-7* have found substances with the biologic properties of siderophores in mold-ripened cheese, such as the Camembert and blue varieties (S. Ong, personal communication), and in soil and dung.[503] The presence of these substances in excrement accounts for the natural growth of the coprophylic fungus, *Pilobolus*.[372] Several members of the ferrichrome family of siderophores have been detected in the Japanese fermented beverage, sake, which is prepared with the aid of *Aspergillus oryzae*. The specific siderophore schizokinen occurs in the blue-green microalgal species *Anabaena*[717] and the capacity to synthesize hydroxamate chelates may enable such organisms to monopolize the limited iron supply in lake waters and so to dominate other algal forms.[573]

Pharmacology   Aromatic hydroxamic acids are potent carcinogens, but this property has not been ascribed to the aliphatic members of the series. If they hydroxamic acid bond were hydrolyzed, an alkyl-substituted hydroxylamine would result. Hydroxylamines are well known to be mutagenic and therefore may be carcinogenic.[135] Little is known about the pharmacology of this class of microbial products, although rhodotorulic acid has been touted as a possible drug for treating persons suffering from chronic iron overload.[322] Desferal, the mesylate salt of deferriferrioxamine B, is the drug of choice for treating acute episodes of accidental iron poisoning in small children.[24] Neither the hydroxamate ligands nor their iron complexes appear to be very well absorbed by humans. Hence, the substances must be injected to treat chronic iron overload. To control acute iron poisoning, the agents need to be administered orally as well as injected.

TOXICITY

Iron- and Sulfur-Oxidizing Bacteria

On the prebiotic earth, the elements were in their reduced states. They still retain this form in those segments of the lithosphere not contacted by the oxygen gas derived from photosynthesis.[94] Exposure of inorganic sulfur compounds, such as iron pyrite, to oxygen will result in the oxidation of the sulfur and the generation of acid. The latter, in turn, will stabilize the ferrous ion and maintain it in solution as a substrate for the "iron bacteria." These reactions lead to acid mine water and such low pH values that few forms of life other than the *Thiobacilli* can survive. The reactions bring about the precipitation of iron, as exemplified by the gelatinous, golden-yellow ferric oxyhydroxide.

Acid mine water is the product of a complex series of transformations of iron pyrite and related sulfide minerals. Some of these transformations are chemical and some are catalyzed by specialized microbes.[352, 510, 873] Pyrite and marcasite (both FeS_2) are oxidized in air to a solution of ferrous sulfate and sulfuric acid:

$$FeS_2 + H_2O + 3\tfrac{1}{2} O_2 \rightleftharpoons FeSO_4 + H_2SO_4$$

The autotrophic bacterium, *Thiobacillus ferrooxidans*, utilizes the ferrous ion as an energy substrate:

$$Fe^{2+} + \tfrac{1}{4} O_2 + H^+ \xrightarrow{\textit{T. ferrooxidans}} Fe^{3+} + \tfrac{1}{2} H_2O$$

Pyrite reduces the ferric ion, precipitating elemental sulfur (S^0):

$$2Fe^{3+} + FeS_2 \rightleftharpoons 3Fe^{2+} + 2S^0$$

The elemental sulfur may be oxidized by the ferric ion or by the sulfur-oxidizing autotroph, *Thiobacillus thiooxidans*:

$$2S^0 + 12Fe^{3+} + 8H_2O \rightleftharpoons 12Fe^{2+} + 2SO_4^- + 16H^+$$

$$S^0 + 1\tfrac{1}{2} O_2 + H_2O \xrightarrow{\text{\textit{T. ferrooxidans}}} 2H^+ + SO_4^-$$

Although the rate-limiting step in this series needs to be defined, the process can be retarded by covering the exposed minerals or eliminating standing water.

Iron and Infection

Because pathogenic microbes require iron and because available iron in host tissues may be limiting, organisms with a well-developed system for acquiring this metal may have special virulence. This subject has been reviewed by Weinberg,[829] who believes that the capacity to synthesize siderophores is a prime determinant of virulence. This thesis is supported by his observations that: elevated temperatures in the host (fever) inhibit synthesis of specific siderophores (e.g., enterobactin);[306] susceptibility to infections is enhanced in hyperferremic animals and in hypotransferremic individuals; and devious mechanisms are employed by the host to deny iron to invading pathogens. Among the infections claimed to be dependent on iron supply are gram-negative septicemia and meningitis, malaria, coliform pyelonephiritis, gas gangrene, listeriosis and systemic candidiasis. Payne and Finkelstein[619] have demonstrated a role of iron in the virulence of *Neisseria gonorrhoeae*.

As noted, factors imparting nutritional immunity have been termed pacifarins, one of which, in the case of salmonellosis in the mouse, has been characterized as enterobactin.[825]

In surveying the role of iron in infection and immunity, Bullen et al.[137] concluded that, taken as a whole, there was "ample evidence" that iron is of crucial importance "in the mechanism of resistance to a variety of bacterial infections." Yet Hegenauer and Saltman[356] have argued that in many instances the observed growth-promoting effect of highly saturated iron transferrin may be attributed to release of metal ion bound adventitiously to nonspecific sites on the protein. Sussman[750] reviewed the relevant data and concluded that, despite the evidence from laboratory animals, hyperferremia does not contribute to the course of human infections. Miles and Khimji[548] found no correlation between pathogenicity and capacity of enteric bacteria to synthesize chelates.

Yancy et al.[869] showed that a strain of *Salmonella typhimurium* rendered defective in the biosynthesis of enterobactin displayed greatly reduced virulence in the mouse. However, the debate still continues. Ultimately, the concept of "optimal" iron nutrition, as noted by Chandra,[160] may prove to be correct.

Bacterial Toxins

Although the formation of a number of bacterial exotoxins is known to be regulated by iron, diphtheria toxin is of special interest in view of the advances in knowledge of its biochemistry and molecular biology.[573] The toxin acts catalytically to effect the adenosinediphosphorylribosylation of the EF-2 protein required for polypeptide chain elongation in the translocation process. Toxin production is programmed by a β-DNA phage that infects *Corynebacterium diphtheriae*. In the presence of bacterial extracts, iron binds the β-phage of DNA to nitrocellulose filters, suggesting that iron is a corepressor of the *tox* gene.[571]

Mechanisms triggering dinoflagellate blooms, those "red tides" that cause mass mortality of fish and other aquatic life, are as yet incompletely understood. One factor triggering their activity may be the availability of iron[443] (cf. chapter 1).

MICROBIAL CORROSION

Microbial corrosion of iron has been reviewed thoroughly by Iverson.[407] Economic losses attributed to corrosion of iron and steel in the United States have been estimated to be from $500 million to $2 billion annually. Corrosion, defined as "the destructive attack of a metal by chemical or electrochemical reactions within its environment," includes rusting, tarnishing, patina formation, pitting, selective leaching, stress cracking, and intergranular corrosion. Both fungi and bacteria are involved in corrosion processes. The most active bacteria are *Thiobacillus*, *Desulfovibrio*, and *Desulfotomaculum*, all of which transform sulfur compounds. By virtue of the elaboration of oxygen gas, algae are also involved. The chemical agents involved in biologic corrosion are varied, and include inorganic and organic acids, hydrogen sulfide, elemental sulfur, mercaptans, and other substances. Microorganisms growing on the surface or in the vicinity of metals may produce substances that establish electrochemical cells leading to corrosion. Biologic corrosion can be combatted by altering the environment in some way, such as by using microbial inhibitors, protective coatings, or cathodic protection.

3
Iron and Plants

Iron is essential to the growth processes of all plants. If a plant is green, it usually has adequate iron. Green plants require a continuous supply of iron as they grow, because iron does not move from older to newer leaves.[113, 121] Iron absorption and transport are genetically controlled by the rootstock.[119, 121] For example, when the iron-efficient Hawkeye soybean top (*Glycine max* L., Merr.) is grafted to the iron-inefficient PI-54619-5-1 (T203) soybean rootstock, iron chlorosis, shown in Figure 3-1, develops in the youngest leaf because the rootstock cannot make iron available from the soil.[121]

Soils do not usually lack iron *per se*,[455] but in most calcareous soils availability of ferric ions may be insufficient for plant growth.[606] For each pH unit increase above 4.0, the solubility of Fe(III) decreases by a factor of 1,000.[469, 812] When the amount of iron available to plants does not meet their minimum needs, plants develop chlorosis, a mineral deficiency disease that will manifest itself in the yellowing or blanching of normally green parts, such as leaf tops, and may cause the plants to die.

Chlorosis is more common in plants grown in alkaline soils. However, some plants grow well on alkaline soils because they are endowed with a biochemical mechanism that makes iron available to them from the soil, that is, they are iron-efficient. The plant and the soil must be compatible for maximum efficiency, and the factors involved in such a relationship are discussed in this chapter for representative species.

In the United States, iron deficiency is most likely to occur west of 100° longitude (roughly the western half of the country). Lock[504] found approximately 258 species or varieties of plants in western states exhibiting naturally occurring iron chlorosis. Thorne and Peterson[770] estimated that 55% of the world land area receives fewer than 51 cm of rain annually, and Wallace and Lunt[812] indicated that 25–30% of the world's land surface is calcareous. Hence, iron deficiency in plants is most likely to be found in these areas and is considered a world problem.

IRON SUPPLY

Inorganic and Stored Iron

The most common sources of iron for plants are the seed itself, the growth medium, and sprays. Germinating seeds usually contain sufficient

Figure 3-1.   Rootstocks of T203 and Hawkeye soybeans affect the development of iron chlorosis and absorption and translocation of radioiron for Quinlan soil (pH 7.5). Top (left to right) photograph and autoradiograph are of T203 top on Hawkeye rootstock; bottom photograph (left to right) shows Hawkeye top on T203 rootstock. The new leaves of T203 are green on the Hawkeye rootstock, whereas the new leaf of the Hawkeye soybean is chlorotic on the T203 rootstock. No radioiron moved from the old leaf to the new one (see also Figure 3-3). Reproduced by Brown *et al.*[121]

iron to meet nutrient requirements of seedlings.[113] Hyde *et al.*[398] suggested that phytoferritin in cotyledons of the pea was the form of iron stored by young seedlings. Chemical factors that interfere with uptake of iron from the growth medium, such as high pH and phosphate, do not interfere with the plant's use of iron from cotyledons.[19] Most iron added as a spray remains in the plant tissue where it is applied (see Figure 3-2), making it necessary for iron to be applied continually to the new growth.

Figure 3-2.   When the chelating agent FeDTPA was sprayed on chlorotic leaves, it corrected the chlorosis only in spots where the residue accumulated (right specimen). Note that the new leaf (left specimen) that developed after the plants were sprayed is completely chlorotic. Reproduced from Holmes and Brown.[388]

Iron nutrition in plants depends largely on chemical factors in roots that affect absorption and translocation of iron from the growth medium, e.g., release of hydrogen ions and reduction of ferric iron. It is not uncommon to see two varieties of the same species growing in the same alkaline soil, with one variety iron-deficient (chlorotic) and the other iron-sufficient (green), as shown in Figure 3-3. The soybean cultivars of Figure 3-3 differ in their ability to respond to iron stress.[117] The chemical

Figure 3-3. Plant species (bottom photograph) and varieties within species (top photograph) differ in their susceptibility to iron deficiency. Top (left to right) T203 (iron-inefficient) and Hawkeye (iron-efficient) soybeans grown on Quinlan soil (pH 7.5). The iron-inefficient plants serve as indicators. Photographs courtesy of J. C. Brown.

reactions that release hydrogen ions and reduce ferric iron are induced by iron stress within the plant where the roots and soil meet and make iron available to the plant. Plants are classed iron-efficient if they respond to iron stress and iron-inefficient if they do not. Iron-inefficient plants become chlorotic and will die on many alkaline soils unless iron is made soluble by chelation, as illustrated in Figure 3-4.

Chelated Iron

Iron chlorosis can be corrected by adding the appropriate iron chelate to the soil or nutrient solution.[159, 355, 388, 423, 486, 813] The primary role of the chelating agent is to make the iron water-soluble and more accessible to the plant root, because roots extract iron from synthetic chelates,[776] and ferric chelates must be reduced to ferrous ones before the ferrous ion can be absorbed by the plants.[161] For a more detailed discussion of the subject, see Tiffin.[775]

Plant species differ in ability to absorb iron from iron chelates, and chelating agents themselves can compete with roots for iron.[129] The ability of EDDHA (ethylene diamine di o-hydroxyphenylacetic acid) to chelate iron was determined in nutrient solutions containing ethylenediaminetetraacetic acid (EDTA), diethylenetriaminepentaacetic acid (DTPA), and cyclohexanediaminetetraacetic acid (CDTA), as competing chelating agents.[129] The apparent stability constants for FeEDTA, FeCDTA, FeDTPA, and FeEDDHA are 24.8, 29.3,[82] 27.9, and greater than 30,[297] respectively. When chelating agents and iron are in equal molar concentration, EDDHA competes successfully with EDTA, DTPA, and CDTA for iron.[129] However, when the concentrations of chelating agents increase, they compete with EDDHA for iron in the following order: EDTA < DTPA < CDTA,[129] a relationship diagrammed in Figure 3-5.

Plant roots differ, as do chelating agents, in their ability to compete for iron.[129, 130] Wheatland milo (*Sorghum bicolor* L., Moench) was unable to use iron from FeEDDHA unless the iron concentration greatly exceeded the EDDHA concentration. For example, at 2×10^{-5}M iron and 0.16×10^{-5}M EDDHA, the milo took up 120 μg iron. At the same iron concentration (2×10^{-5}M), but with the EDDHA concentration elevated to 1×10^{-5}M, the sorghum took up only 13 μg iron.[113] As can be seen in Figure 3-6, okra (*Hibiscus esculentus* L.) and wheat (*Triticum aestivum* L.) developed iron-deficiency chlorosis when the molar concentration of DTPA exceed the molar concentration of iron.[113] Plant species may be able to alter the activity of a metal ion by increasing or decreasing the concentration of a specific chelating agent inside the plant or in the root exudate. In this way, the type and concentration of a chelating agent, coupled with iron availability, determine the uptake of iron and other nutrient elements by plants. Both the growth medium and the

Figure 3-4. Iron-inefficient T203 soybeans grown on 16 calcareous soils (left four plants) responded to 150 kg FeDTPA/ha added to their soils (right four plants). Photograph courtesy of J. C. Brown.

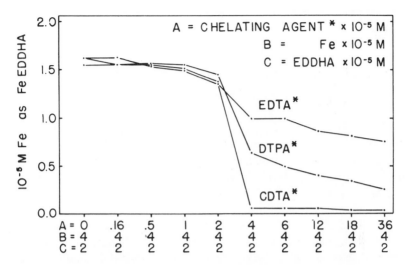

Figure 3-5. The capacity of EDDHA to form FeEDDHA was affected by the concentration of the competing chelating agents in solution at 4×10^{-5}M iron. No competition for iron existed until the molar concentration of chelating agents exceeded the molar concentration of iron. Reproduced from Brown et al.[129]

Figure 3-6. When the molar concentration of DTPA exceeded the molar concentration of iron, DTPA competed with the roots of okra and wheat (top and bottom photographs, respectively) for iron. The nutrient solution contained 1×10^{-5}M iron and (both genotypes, left to right) 0.16, 1, 2, 6, and 18×10^{-5}M DTPA. Reproduced from Brown et al.[130]

plant variety itself contribute to making a continuous supply of iron available.

CONDITIONS AFFECTING IRON UPTAKE AND TRANSPORT

Soil and Environment

The hydrogen ion concentration of a soil is a crucial factor governing the distribution of plants in nature. Some plants grow well only on acid soils; others grow better on alkaline soils. For example, bentgrass (*Deschampsia flexuosa*) develops iron-deficiency symptoms in a growth medium with a pH above 6,[608] whereas mustard (*Sinapis alba* L.) grows well under similar conditions.

Wallace and Lunt[812] listed the following as principal causes of iron chlorosis: poor iron supply; excessive calcium carbonate; bicarbonate in soil or irrigation water; overly irrigated or high-water conditions; high phosphate; high concentrations of heavy metals such as manganese, copper, and zinc; low or high temperatures; high light intensities; high concentrations of nitrate nitrogen; unbalanced cation ratios; poor aeration; certain organic additions to the soil; viruses; and root damage by nematodes or other organisms. All these factors are more effective in a natural alkaline soil, and several may be operating at the same time. For example, the role of bicarbonate in precipitating iron depends, in part, on the interrelationship between bicarbonate, phosphate, calcium, and ferric ions in the soil. A moist calcareous soil containing decomposing organic matter provides a condition for maximum bicarbonate-ion accumulation that may increase phosphate availability and decrease the iron available to plants.[127] Bicarbonate *per se* does not appear to be a direct cause of iron-deficiency chlorosis.[112] Instances of chlorosis have been associated more with phosphate concentration in solution than with bicarbonate anion concentration.[112] Phosphate is considered a ligand that competes with the plant for iron.[529] Greenwald[327] and Olsen *et al.*[610] observed that bicarbonate increased the solubility of phosphorus in solution. Phosphate strongly influences iron absorption and translocation in plants.[73] This phenomenon has been well characterized.[113, 568, 771, 812] More recent research[117] has indicated that use of iron by plants is dependent on the plant species or variety grown. Plant nutrition is entering an era in which equal emphasis is rightly being placed on the plant as well as on the soil for improving the efficiency of crop production.

Plant Response to Iron Stress

The term "iron stress" implies that a plant is deficient in iron. Plants are classed as iron-efficient if they respond to iron stress by inducing biochemical reactions that make iron available in a useful form, and iron-inefficient if they do not. An iron-efficient plant may respond to iron stress without any visual iron-deficiency symptoms, whereas an iron-inefficient plant develops chlorosis. When plants respond to iron stress, the following products or biochemical reactions are more likely to occur in iron-efficient than in iron-inefficient plants:

Release of hydrogen ions from the roots
Release of reducing agents from the roots
Reduction of ferric iron at the roots
Increases in organic acids (particularly citrate) in roots

Response to iron stress is adaptive and is known to be determined genetically in several plant species.[55, 816, 835]

Hydrogen ions released from roots Kirkby and Mengel,[446] working with tomatoes (*Lycopersicon esculentum* Mill.), reported an elevated pH in nutrient solution with nitrate, and a reduced pH with ammonium nutrition. With nitrate as the source of nitrogen, the pH increased to pH 7.2 for T3238fer (iron-inefficient) tomatoes but decreased to pH 4.3 for T3238FER (iron-efficient) tomatoes when the plants were subjected to iron stress.[125] When supplemental iron was added to the iron-sufficient T3238FER tomatoes, hydrogen ions were not released from their roots and the pH increased to 7.2, the same as for T3238fer. During an 8-hr absorption period, iron-stressed T3238FER plants absorbed and translocated approximately 20 times more iron-59 to their tops than iron-sufficient T3238FER, and 78 times more iron-59 than iron-stressed T3238fer tomatoes.[125] The iron-inefficient T3238fer did not respond to iron stress under any of the experimental conditions.

When hydrogen ions are released from roots and the pH is lowered to make the environment of the root more acidic, iron is made more available for plant uptake.[608] Although beneficial, this change is not always the ultimate solution to iron chlorosis. For example, iron-stressed Wheatland and "B-line" (iron-inefficient) sorghum and Pioneer 846 and KS5 (iron-efficient) sorghum release about the same quantity of hydrogen ions into the growth medium, but they differ in their absorption and transport of iron.[123, 126] This difference is associated with greater reduction of Fe(III) to Fe(II) at the root[126] by the iron-efficient than by the iron-inefficient sorghum lines. Each factor has its own effect on how plants use iron and may differ with variety or plant species and their growth environment.

Reducing agents released from roots In addition to releasing hydrogen ions, iron-efficient soybeans[122] and tomatoes[119] release "reductants" from their roots in response to iron stress. "Reductants" is a word coined to designate compounds released by roots that reduce

Fe(III) to Fe(II). Chelating agents interfere with the activities of reducing agents, but such interference may be eliminated by increasing the concentration of the reductant in solution.[116]

Iron-stressed Hawkeye (iron-efficient) soybeans released more reductant into solution than the iron-stressed T203 (iron-inefficient) variety, but iron uptake was not increased when the T203 plants were placed in the Hawkeye solutions.[116] This may mean that reductants in the external solution indicate a leaky root resulting from the release of hydrogen ions into the nutrient solution, although reductants have been found in nutrient solutions at pH 7.[126] More important may be the adaptive production of reductants inside the root or at the root surface that keeps iron in the more available ferrous form.[20, 115]

Over the past 20 years, reducing agents associated with iron transport have been identified in microorganisms. The excretion of metal-binding phenolic acids by iron-stressed *Bacillus subtilis* might be a mechanism to correct iron deficiency.[406] Phenolic acids accumulated by iron-stressed *Bacillus subtilis* do not seem to be involved in iron uptake, but serve to solubilize the metal in the growth medium.[628] The addition of iron to growth cultures will inhibit phenolic acid excretion by *Bacillus subtilis*.[142] Although not identified, the reductants released by iron-efficient soybeans[20] and tomato plants[115] respond similarly to the phenolic acids released by *Bacillus subtilis*.

Ferric iron reduced at root Reduction of Fe(III) to Fe(II) is another factor induced by iron stress and it occurs principally in the young lateral roots of iron-efficient soybeans[20, 116] and tomato plants.[115] Sites of reduction were determined by transferring the iron-stressed plants to nutrient solutions containing FeHEDTA and potassium ferricyanide, $K_3Fe(CN)_6$.[115] A blue precipitate (referred to here as Prussian blue) developed in the epidermal areas of the root where one form of the ferric ions was reduced by the root, shown in Figure 3-7. Reduction of Fe(III) occurred in areas that were outside the root and accessible to bathophenanthrolinedisulfonate (BPDS).[115] Reducing conditions were established by adding BPDS, 10% in excess of Fe(III), to the nutrient solutions. As the ferric ion was reduced, Fe(II) was trapped in solution as ferrous $BPDS_3$. Hence most of the iron was not transported to the plant top.[115, 118]

If the iron-efficient soybean roots were given iron as FeHEDTA for 20 hr and then rinsed free of FeHEDTA and placed in nutrient solutions containing potassium ferricyanide, Prussian blue (indicative of ferrous iron) appeared throughout the protoxylem of the young lateral roots, as can be discerned in Figure 3-8, and throughout the regions where the root elongates and matures in the primary root[20, 114] (see also Figure 3-7, left). According to this test, the metaxylem of the iron-efficient plants contained no ferrous ions.

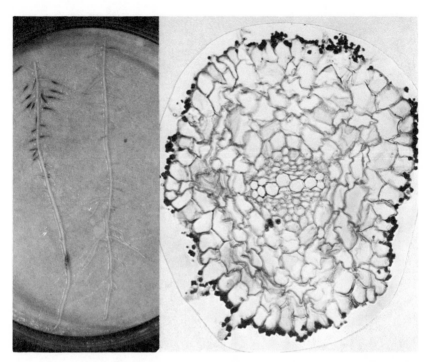

Figure 3-7. Left photograph shows sites of reduction of Fe(III) to Fe(II) (dark areas are Prussian blue formations) on lateral roots, and on the area of elongation and maturation of the primary root of T3238FER (iron-efficient) tomatoes. No reduction can be found on the T3238fer roots. The right photograph is a cross section of a lateral root of T3238FER. Prussian blue crystals have formed on the periphery of the root. The three dark spots inside the root reveal contamination occurring while the specimen was being photographed. Reproduced from Brown and Ambler.[115]

Evaluation of organic acids (particularly citrate) in roots Iron-deficient plants usually contain more citric and malic acids than normal green plants.[233, 400, 401, 664] Citric acids chelate iron and keep it soluble in an external solution, and they may function similarly inside the plant.[671] A striking relationship was observed to exist between iron and citrate transported in the xylem exudate,[118, 128] which is plotted in Figure 3-9. When iron increased, the citrate increased; a decrease in iron was paralleled by a decrease in citrate. This relationship held if iron stress were induced in the plant by limiting the iron supply or if zinc, azides, or arsenate[118, 128] were used to induce iron stress. Tiffin,[772, 774] using electrophoresis to follow the migration of the chelated metal, identified ferric citrate in the xylem exudate of several plant species. Enough citrate was always present to chelate the metal, and citrate in excess of that needed for iron chelation migrated as an iron-free fraction behind the iron-citrate band.

Clark et al.[176] showed that malic, acetic, and *trans*-aconitic acids

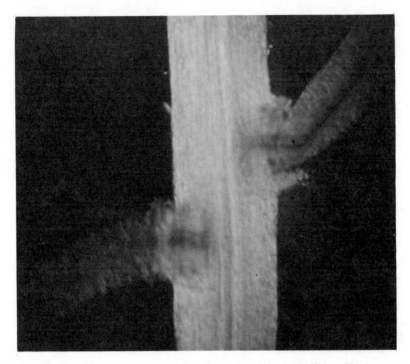

Figure 3-8. Protoxylem of iron-stressed lateral roots of the iron-efficient Hawkeye soybean contained Prussian blue (dark areas) indicating Fe(II) continuous in protoxylem until union with metaxylem of the larger root, enlarged 20 times. The iron-stress-response mechanism reduced Fe(III) to Fe(II) and kept the ferrous ions available in the protoxylem. Reproduced from Ambler et al.[20]

were ineffective in moving iron-59 electrophoretically in acetate, citrate, isocitrate, *trans*-aconitate, and malate buffers. Citric acid moved iron to the annode whenever present on the electrophoretogram, and competed successfully with the other acids for iron. Clark *et al.*[176] also determined that iron-efficient WF9 corn (*Zea mays* L.) absorbed and transported more iron than iron-inefficient ys_1 corn, but the latter contained sufficient citric acid in the xylem exudate to transport iron *in vitro*. These findings indicate that the iron-inefficient corn roots do not respond to iron stress and the metal is not made available for transport in the xylem exudate. The translocation of iron in the plant involves more than citrate chelation of iron in the root *per se*.

ROLE OF THE IRON-STRESS-RESPONSE MECHANISM

Internal Root Control

The term "iron-stress-response mechanism" is used to denote a positive response to iron deficiency that induces biochemical or physiological reactions within the plant that make iron available for plant use. This mechanism operates in green plants that grow well on calcareous alkaline soils. Chlorotic plants found in such land areas are iron-deficient because the iron-stress-response mechanism is not functioning. Clark and Brown[175] found that when WF9 (iron-efficient) and ys_1 (iron-inefficient) corn genotypes were grown together, with ys_1:WF9 ratios of 4:0, 3:1, 2:2, 1:3, and 0:4 plants per container, WF9 in the presence of ys_1 released more hydrogen ions, reduced more Fe(III) at the root surface, and took up more iron than the ys_1 genotype. All ys_1 plants, regardless of the ratio to WF9, developed iron chlorosis.[175] In a similar study using Hawkeye and T203 soybeans, with T203:Hawkeye ratios of 28:0, 24:4, 16:12, 4:24, and 0:24, Ambler and Brown[18] found that, regardless of ratio, Hawkeyes took up 80% more iron than T203's. In another study,[116]

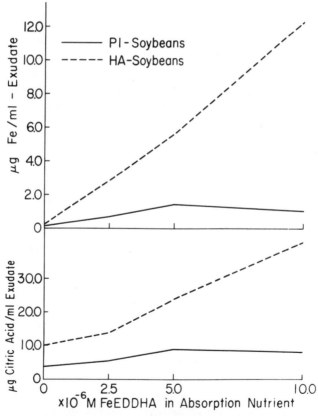

Figure 3-9.   When the iron concentration in the stem exudate was increased by increasing the iron concentration in the nutrient solution, the citrate concentration in the exudate also increased in Hawkeye (HA) soybeans, but not in the iron-inefficient T203 variety. Both cultivars were subjected to iron stress before being transferred to the nutrient solutions containing different concentrations of the metal. Reproduced from Brown and Tiffin.[128]

iron uptake by T203 soybeans was not increased when they were placed in solutions in which iron-stressed Hawkeye soybeans had released hydrogen ions and reducing agents and had absorbed and transported 80% of the iron available. It was concluded that iron absorption and transport are controlled inside the roots and iron uptake is greatest when the response mechanism is functioning.

Both iron-efficient and iron-inefficient roots can have several hundred micrograms of iron per gram of root, but the iron-inefficient plant may die from lack of iron in its tops. In contrast, iron is made available to tops by the iron-efficient roots in response to iron stress.[123] In a similar way, iron may remain in the nutrient solution as a ferric chelate[119] or a ferric phosphate,[114] and not be taken into the plant until it is made available for absorption and transport through chemical reactions induced by iron stress. Iron usually is used in plant tops once it is made available for transport by the roots.

Mechanism of Iron Uptake

Iron absorption and transport, induced in response to iron stress, involves the release of hydrogen ions by the root, which lowers the pH at the root zone. This favors $Fe(III)$ solubility and reduction of $Fe(III)$ to $Fe(II)$. Iron-efficient roots also release reductants in response to iron stress. These agents, along with reduction at the root surface, reduce $Fe(III)$ to $Fe(II)$, which enters the root primarily through the young lateral roots. It is likely that the $Fe(II)$ is kept reduced in the roots by the reductant. Ferrous ions are present throughout the protoplasm and may or may not have entered the root by a carrier mechanism. The root-absorbed $Fe(II)$ is oxidized to $Fe(III)$ near the metaxylem, chelated by citrate, and transported in the metaxylem to the top of the plant.

Other Chemical Reactions Affected

The chemical reactions induced by iron deficiency may affect a plant's nitrate reductase activity, its use of iron from ferric phosphate and from FeEDDHA, and its tolerance to heavy metals.

Nitrate reductase activity The products of iron stress concomitantly increase nitrate reductase activity in roots, shown in Figure 3-10 for tomatoes. In both induced nitrate reductase activity and induced "reductants" activated in response to iron stress, a substrate is reduced: i.e., nitrate to nitrite ions by nitrate reductase,[735] and $Fe(III)$ to $Fe(II)$ by the reductant.[115] The induced nitrate reductase activity decreased when iron was made available to the plants.

Use of iron from ferric phosphate[159] Phosphate, the plant, and the chelating agent [ferrous ferrozine, Fe(II)3-(2-pyridyl)-5,6-*bis*(4-phenyl-sulfonic acid)-1,2,4,triazine], compete for the iron in nutrient solution. Hawkeye and T203 soybeans were used as test plants. Within 2 days

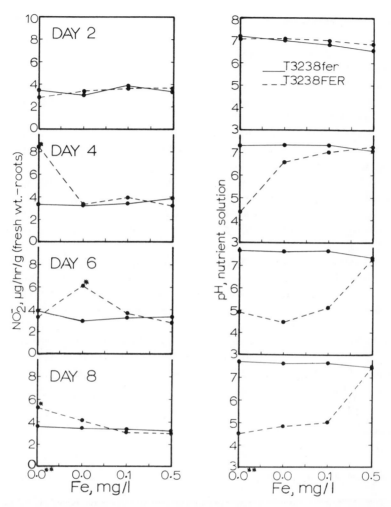

Figure 3-10. When subjected to different iron concentrations in nutrient solutions (NO$_3$-N), nitrate reductase activity increased in T3238FER roots, and the pH decreased in the nutrient solution in response to iron stress. Neither the iron-sufficient T3238FER nor the T3238fer tomato plants showed these responses. The plants were 21 days old when transferred to these treatments. *Statistically significant at 1% level. **Some iron removed from roots with EDDHA treatment. Drawings courtesy of J. C. Brown and W. E. Jones.

after adding 16 mg phosphorus as phosphate to the nutrient solution, only about 10% of the ferrous ferrozine was colorimetrically detected in solution. The iron was removed from ferrous ferrozine, and it appeared as a suspension of iron phosphate. Phosphate was dominating the system for iron. Four days later, all plants developed some chlorosis in the new leaves. The iron-efficient soybeans responded to this stress by releasing more hydrogen into solution and reducing more ferric to ferrous ions than the iron-inefficient variety. During this process, ferrous ferrozine reappeared in the solutions containing the Hawkeye plants, but not the T203 specimens. The iron-efficient plants now dominated the competition for iron. On day 17 (final harvest), the green iron-efficient soybeans contained 86 μg iron/g of tops, whereas the chlorotic iron-inefficient soybeans contained only 36 μg iron/g of tops. In the latter, phosphate still dominated the system for iron.

Use of iron from FeEDDHA[119]   Iron-inefficient T3238fer tomato plants developed iron chlorosis because they could not absorb iron from FeEDDHA, as shown in Figure 3-11. In contrast, the iron was available to iron-efficient T3238FER tomato plants because they could reduce Fe(III) to Fe(II). In order for plants to use Fe(III) from several ferric chelates, they first must reduce the ferric chelates to ferrous chelates,[161] which generally have a much lower stability constant than the ferric compound.

Tolerance to heavy metals[124]   Heavy metals added to soils in pesticides, fertilizers, manures, sewage sludge, and mine wastes can cause various degrees (depending on genotype) of iron deficiency to develop in plants. An iron-efficient plant is more tolerant of heavy metals than an iron-inefficient one. For example, copper added as a fungal spray to citrus trees and zinc added as the oxysulfate to control bacterial spot in peach trees accumulated in the soils in concentrations toxic to plant growth.[124] When iron is made available, it counteracts the effect of the heavy metals.[726]

GENOTYPIC DIFFERENCES AND THEIR RELATION TO IRON STRESS

Iron Deficiency and Excess

A recessive gene controls the uptake of iron in iron-inefficient T203 soybeans,[835] ys_1/ys_1 corn,[48, 55] and T3238fer tomato plants.[816] These plants usually develop iron chlorosis on alkaline soils, and to a much lesser extent on some acid soils. In contrast, the most iron-efficient plants are usually green on alkaline soils, and may even develop symptoms of iron poisoning on some acid soils. For example, Olsen[608] found that bentgrass developed iron-deficiency symptoms in nutrient solutions (pH above 6), but mustard grew well under these conditions with ferric sulfate as the source of the metal. When the pH of these solutions decreased from 6 to 4, symptoms of iron toxicity developed in the mustard, but not the bentgrass. These same relationships were found in soils.[608] Iron-efficient Bragg soybean tops were reported to contain 45, 353 and 1,320 μg iron/g and iron-inefficient Forrest soybean tops had 22, 54, and 112 μg iron/g when grown on alkaline Quinlan soil (pH 7.5), acid Bladen soil (pH 4.3),

Figure 3-11.   T3238fer (left) and T3238FER (right) tomato plants as they appeared when grown in nutrient solution containing 10 μM iron supplied as FeEDDHA. The iron-inefficient T3238fer tomato could not use iron from FeEDDHA. Reproduced from Brown et al.[119]

and in nutrient solution containing 2 mg iron/liter as FeHEDTA, respectively (unpublished observations, J. C. Brown and W. E. Jones). Bragg soybeans showed signs of iron overload in the nutrient solution and Forrest soybeans developed symptoms of iron deficiency on the Quinlan soil. These relationships are illustrated by Figure 3-12. Iron toxicity is only about 10% as prevalent as iron deficiency. However, the increased uptake of the metal by iron-efficient plants grown on acid soils may be injurious to plant growth, although the condition has not been recognized or documented.

Iron and its Relationship to
Phosphorus-Efficient and Phosphorus-Inefficient Plants

Mikesell *et al.*[547] concluded that B-line sorghum was less iron-efficient than KS5 line sorghum because the B-line took up more phosphorus than the KS5 line and that phosphorus interfered with the availability of iron. Brown and Jones[126] confirmed these findings and showed further that the KS5 line responded more to iron stress than B-line. They concluded that the accumulation of phosphorus and insufficient iron-stress response both contributed to iron deficiency in B-line sorghum.

a   b   c   d

Figure 3-12.   Bragg (a) and Forrest (b) soybeans grown in nutrient solutions containing 2 mg iron/liter and grown (c and d) in Quinlan soil (pH 7.5). Note the symptoms of excessive iron on Bragg (a), but not on Forrest (b), and the iron-deficiency symptoms on Forrest (d), but not on Bragg (c). Photographs courtesy of J. C. Brown and W. E. Jones.

Figure 3-13.   When SC369-3-1JB, PI-405107, and NK212 (top left to right) sorghum were grown on Quinlan soil (pH 7.5), SC369-3-1JB developed iron chlorosis, but NK212 did not. In the bottom series of photographs, NK212 specimens grown on Bladen soil (pH 4.3) developed phosphorus-deficiency symptoms, but SC369-3-1JB did not. The PI-405107 genotype was intermediate in its response. Photographs courtesy of J. C. Brown, R. B. Clark, and W. E. Jones.

Other sorghum lines grown on alkaline soil developed iron chlorosis on the phosphorus-efficient but not on the phosphorus-inefficient genotypes (unpublished observations, J. C. Brown and W. E. Jones). When the sorghum lines were under phosphorus stress (deficient in phosphorus), only the phosphorus-efficient genotypes survived, as can be seen in Figure 3-13. The phosphorus-efficient genotypes took up 15 times more phosphorus-32 than the phosphorus-inefficient plants.

Zinc Stress as an Inducer of Iron Uptake

For some unexplained reason, zinc deficiency induces iron uptake in some plant species or varieties.[17, 409, 822] Zinc-deficient corn plants accumulate excessive iron in their tops; evidently the excessive iron concentration was associated with zinc stress in the plant and not with the level of iron in the soil solution.[409, 822] Ambler and Brown,[17] working with Sanilac (zinc-inefficient) and Saginaw (zinc-efficient) navy beans (*Phaseolus vulgaris* L.), showed that when Sanilac plants developed zinc-deficiency symptoms, as in Figure 3-14, they contained twice as much iron (655 μg

Figure 3-14.   Zinc-deficiency symptoms appear on Sanilac leaves (left) but not on Saginaw plant leaves (right), when both are grown in a split medium of Shano soil (top) and nutrient solution (bottom). Reproduced from Ambler and Brown.[17]

iron/g) and nearly twice as much phosphorus (1.7%) in their tops as Saginaw. Sanilac tops contained 18 μg zinc/g compared to 23 μg zinc/g in Saginaw tops. Zinc-deficient cotton (*Gossypium hirsutum* L.) contained 412 μg iron/g compared to 94 μg iron/g top in the zinc-adequate plants (unpublished observations, J. C. Brown and W. E. Jones).

Interference of Molybdenum Stress with Iron Uptake

Molybdenum is required by flavoproteins that transfer electrons[375] and by nitrate reductase, the enzyme that breaks down nitrate.[53, 237, 271, 594] Barry and Reisenauer[63] determined that molybdenum affected the reductive capacity of Marglobe tomato roots. Iron uptake was depressed when molybdenum concentrations were both low (less than adequate) and high (more than adequate) in the nutrient solution. The plants receiving adequate molybdenum were the most effective in raising the redox potential, and those that were not supplemented were the least effective. The increased redox potential corresponded to the production of ferrocyanide, i.e., the reduction of Fe(III) to Fe(II), and an increased uptake of iron by the plant. Iron accumulation was maximal at marginally adequate levels of molybdenum nutrition.

Role for Plant Breeding

The plant breeder must select or develop iron-efficient cultivars to be used where iron deficiency is a problem in crop production. In past years, and with very little economic success, attempts have been made to correct problems of iron nutrition by changing the soil to fit the plant. Plant breeding now offers the possibility of controlling iron chlorosis by tailoring the plant to fit a problem soil. In some cases, varieties, lines, or hybrids already developed or selected will overcome the problem. In others, however, more iron-efficient lines will need to be developed.

Citrus[867] and grape[817] growers in Texas and Utah, respectively, controlled iron chlorosis genetically by grafting iron-efficient rootstocks onto desirable scions. Sprague[729] believes that genetic knowledge is adequate to provide the necessary support for a productive cooperative effort between physiologists and geneticists to develop iron-efficient plants. Heslop-Harrison suggests that "the most effective way to obtain a growth pattern efficient in a given environment for a particular purpose is to breed a genotype for the job."[370] It is necessary to know the nutrient requirements of crop plants before fitting them to a particular soil.[4] For example, when the Hawkeye soybean was replaced by new, iron-inefficient soybean varieties in central and north-central Iowa (calcareous soil), iron chlorosis developed in the new soybean varieties.[237] These plants were probably not tested for iron efficiency before they were released to the field.

FORMS AND FUNCTIONS OF IRON IN PLANTS

If iron absorption and transport depend on the induction of the iron-stress-response mechanism, then these induced biochemical reactions should enhance the use of the element throughout the plant. Iron is an essential component of many heme and nonheme enzymes and proteins.[272, 648, 649] Nucleic acids contain iron,[301] and a ribosomal chromoprotein has been identified in rat liver that contains 20% iron.[546] The iron concentration of this protein varied with the physiologic state of the organism. Storage forms of iron in plants include phytoferritin[398, 593, 700] and ferric phosphate. Iron accumulation in different plant parts generally follows this order: roots > old leaves > young leaves > stems.[784]

Some of the iron in plants is chelated. When a cell develops iron stress, i.e., deficiency, iron chelates with low stabilities will fail to form and the metal will be distributed among other ligands.[648] For example, in catalase, the iron exchange is rapid, and in cytochrome a, it is slow. Catalase activity has been reduced in several plant species as a result of iron deficiency.[120, 832] Price[648] indicates that data are insufficient to decide whether or not iron compounds in physiologic systems behave as simple chelates. He suggests that because chlorosis is one of the earliest symptoms of iron deficiency, the compound responsible for the condition is one of the least stable among the physiologically essential forms of iron. Phosphate may act as a competing ligand for iron in such a system.

This chapter has emphasized iron absorption and transport rather than its forms and functions, which have been thoroughly reviewed.[232, 272, 373, 374, 578, 648, 649]

RESEARCH NEEDS

Plant species and varieties within species differ in their iron requirements and their tolerance to high concentrations of mineral elements, which complicates determining the nutrient requirements of plants so that the plant and the soil can be made compatible. The plant breeder will be challenged to develop plants that are nutritionally adapted to problem soils. The agronomist, plant physiologist, biochemist, and horticulturist will need to supply the geneticists with techniques for identifying a desirable factor or trait for a plant. For example, iron-inefficient plants develop less iron-stress response than iron-efficient plants, and this range in response is the basis of a technique for screening plants for iron-efficiency.[123] A limited supply of iron and some control of pH in the growth medium are required in this technique. Degree of iron chlorosis is all that is needed to give the plants an efficiency rating. When the genetically controlled reaction is identified in the plant that responds to iron stress, a more specific technique may be developed and used by the plant breeder to identify iron efficiency in that particular genotype.

Regional laboratories for testing soils and plant tissues could be established to meet the needs of a specific area. Soils and crops could be characterized to designate which cultivars would be most suitable to the area and what supplements should or should not be added to the soil. The genetic control of nutrition is well established, and a genetic program is suggested to fit plants nutritionally to problem soils, i.e., iron-deficient, saline, and manganese- and aluminum-toxic. Recognition of the role of plants in controlling mineral absorption and transport signals a new direction for plant nutrition. Related programs, based on the pooling of resources and expertise, would conserve our soils, improve the efficiency of fertilizers, and increase crop production.

4

Iron Metabolism in Humans and Other Mammals

TOTAL BODY IRON

The essential concentration of iron in the vertebrate organism increased by two orders of magnitude over that of lower living forms when hemoglobin evolved as an oxygen transport vehicle. Total body iron concentrations of various animal species, estimated on the basis of the sum of individual iron fractions, vary between 25 and 75 mg/kg body weight. The adult human male has approximately 49 mg iron/kg and the adult female has approximately 38 mg/kg, equivalent to a total body iron content for the 80 kg male of about 4 g and about 2.5 g for the 65 kg female.[89] Miscible iron has been determined by isotopic dilution studies to be 42 mg/kg in the male;[323] this somewhat lower estimate is assumed to result from incomplete mixing with the pool of storage iron. Variations in iron content within individuals of one species and among other vertebrate species are explained by differences in circulating hemoglobin mass and/or by differences in storage iron. The general proportions of erythron iron, other essential tissue iron, and storage iron in humans are summarized in Table 4-1.

Essential body iron in humans is generally proportionate to lean body mass at about 35 mg/kg, and a variable amount of storage iron must be added to this amount. At birth, total body iron concentration is increased by the elevated red cell mass and the presence of appreciable storage iron in the liver.[493] During the first month of life, stores increase still further because of the decrease in circulating red cell mass. Thereafter, rapid growth and greater red cell requirements result in a transfer in the opposite direction so that iron stores become virtually exhausted between the sixth and twenty-fourth months.[704, 725] Little direct evidence exists of the size of iron stores in later childhood, but as can be seen in Table 4-2, serum ferritin values indicate a limited iron reserve, perhaps 5 mg/kg until after the age of 15.[192, 666] Iron reserves of some 1,000 mg in the male are created between the ages of 15 and 30; adult females maintain low iron stores of about 300 mg, but after menopause they increase to the level found in males.

Although various chemical measurements of the iron content of body tissues have been made in animals and humans,[787] such analyses are

Table 4-1.   Body iron contents of adult humans

	Male		Female	
	mg/kg	Total body content (mg)[a]	mg/kg	Total body content (mg)[b]
Erythroid marrow	2	160	2	130
Circulating erythrocytes	28	2,240	26	1,690
Mygolobin	≈4	320	≈3	195
Cell enzymes	≈2	160	≈2	130
Ferritin and hemosiderin	13	1,040	5	325
Total	49	4,000	38	2,470

[a] Based on a total body weight of 80 kg.

[b] Based on a total body weight of 65 kg.

of limited utility. The difficulty lies in the lack of attention paid to separating hemoglobin iron from other iron, because of the very large concentrations of iron in circulating red cells that contaminate those tissues. Hemoglobin, ferritin-hemosiderin, and myoglobin are the only quantitatively important body-iron compounds that have been found; therefore, any appreciable concentration of iron in the tissues is assumed to represent one of those components.

Body iron contents of mammals vary with age, depending on the iron endowment at birth, the duration of breast-feeding, the rate of growth, and reproductive requirements.[453, 787] With the exception of the pig, whose exceedingly rapid growth frequently outstrips iron supply,[300] mammals easily acquire and maintain essential body iron, and iron deficiency is exceedingly rare.

THE NATURE OF BODY IRON

Iron Compounds of the Blood

Red cell hemoglobin constitutes the largest fraction of body iron. Hemoglobin is composed of four 16,000-dalton polypeptides, each possessing a heme moiety within which an iron molecule is located. The two different subunits of the protein together form a tetrameric molecule, $(\alpha\beta)_2$. Each chain is coiled into a α-helix folded around the heme group in such a way as to provide a nonpolar environment for its heme. This folding is essential in permitting the iron to combine with oxygen without being oxidized.[626] In unoxygenated hemoglobin, the iron atoms are five-coordinated, and are forced out of the plane of the heme ring.[570] Upon oxygenation, the sixth coordinate site is occupied by oxygen, causing the three electrons of the iron atom to rearrange. This alteration reduces the radius for the iron atom, enabling it to move back into the heme plane. These changes in structure are essential for the normal functioning of hemoglobin as an oxygen carrier.

Hemoglobin accounts for approximately 85% of essential iron in the human female and 60% of the total in the male, a distinction largely related to the difference between the size of iron stores in the two sexes. Hemoglobin concentration in normal humans depends on ambient oxygen tension, hemoglobin affinity for oxygen, and circulating testosterone levels.[3] In childhood, affinity of hemoglobin for oxygen is somewhat decreased, which may be related to an increased organic phosphate concentration. The rise in phosphate is thought to be responsible for a slightly lower hemoglobin concentration because it makes more oxygen available to tissues.[152] The sex difference in adult hemoglobin concentration is undoubtedly linked to the effect of testosterone on erythropoietin stimulation of the marrow.[15] The effect of altitude on hemoglobin is well established. It is proportional to decreases in arterial oxygen saturation, resulting from changes in ambient oxygen tension.[395]

The physiologic norm for an individual's hemoglobin concentration varies considerably, and it is extremely difficult in mild anemia to separate physiologic variations from deviations produced by pathologic states. It has also been complicated to define mean normal values of hemoglobin concentration for normal populations. Surveys are problematic in that they embrace a considerable number of iron-deficient individuals, some of whom are anemic. One study of a Swedish population showed a hemoglobin response of 17% of the "normal" adult population to iron administration.[305]

Statistical approaches have been developed for surveys to exclude individuals with iron deficiency.[188] Mean hemoglobin concentrations from residents of different parts of the world have not demonstrated dif-

Table 4-2. Mean values for serum ferritin as a function of age

Age	Ferritin, μg/liter
Newborn	110
6 months	5
5–10 years	21
12–18 years	22
18–45 male	94
18–45 female	25
>45 male	124
>45 female	89

ferences in hemoglobin concentration based on racial derivation, although the existence of such variances often has been suggested. For example, mean hemoglobin concentrations from individuals in ten states, set forth in Table 4-3, suggest racial differences, but dissimilar dietary customs also may have been responsible. The values in Table 4-3 are certainly influenced by a population with iron deficiency. In subjects sampled from the Pacific Northwest, about 4% of the adult males and 20% of the menstruating females were iron-deficient,[192] and half of that number had demonstrable anemia.

Normal hemoglobin values for other mammals are even less well defined, but reports do show characteristic species variations that influence body iron content.[453, 787] Thus, some sheep have a hemoglobin content approximately half that of a human, apparently explained by a marked increase in oxygen dissociation.[614] In animals such as dogs and goats, hemoglobin concentration may fluctuate markedly because their spleens can temporarily hold as much as one-third of the red cell mass, although the total intravascular hemoglobin in these species is quite similar to that of humans.

Transferrin is the other iron compound of the blood with an important physiologic function.[562] This plasma component is a glycoprotein with a molecular weight of about 80,000; it has a prolated ellipsoid shape with an axial ratio of 1:3 and is composed of a single polypeptide chain with two identical carbohydrate side chains.[6] Although this protein is pleomorphic—with some 18 variants described in humans—these differences are not known to affect iron transport.[311] Transferrin is predominantly synthesized in the liver. Its production and losses parallel those of albumin.[311] The normal concentration of the protein in humans is about 2.3 g/liter of plasma,[188] equivalent to an iron-binding capacity of 3.3 mg/liter. Transferrin concentration changes inversely in relation to changes in body iron stores.[831] In individuals with adequate body iron, the iron-binding capacity of transferrin is only 20–45% saturated; the remainder constitutes a latent capacity. Mean plasma iron concentration and total iron binding capacity in many animal species are quite similar to humans.[89, 787] A higher plasma iron in some animals appears to be a reflection of a much greater amount of iron absorbed—with substitution of a low-iron diet, plasma iron falls to about 100 μg/dl.[193]

Essential Tissue Iron

Three types of iron-containing compounds in the body perform essential metabolic functions.[216] The first type includes all other heme iron compounds. Of these, myoglobin is the largest fraction, amounting to about 15 mg iron/kg muscle.[13] Its structure is closely related to the monomeric unit of hemoglobin: it contains one polypeptide chain attached to a heme group with a single iron atom. It functions as a link in the oxygen

Table 4-3. Median hemoglobin in populations of ten states, 1968–1970[a]

Age	White	Black	Spanish-surnamed
<2	11.1	10.7	12.3
2–5	12.1	11.3	12.5
6–12	12.6	12.0	12.9

Age	Male			Female		
	White	Black	Spanish-surnamed	White	Black	Spanish-surnamed
13–16	14.1	13.0	13.8	13.1	12.3	13.1
17–44	14.9	14.2	15.0	13.3	12.8	12.9
45–59	14.9	13.9	14.6	13.4	12.8	13.4
>59	14.6	13.5	14.2	13.6	12.6	13.4

[a] Derived from data of the U.S. Department of Health, Education, and Welfare.[790]

transport chain, accepting oxygen from the blood and storing it for utilization during muscle contraction. Cytochromes *a*, *b*, and *c* and P450 represent another set of tissue heme proteins located in the mitochondria and other cellular membranes that aid in electron transport.[596] Of these, cytochrome *c*, a pink protein with a molecular weight of 13,000, is the best characterized. Catalase and peroxidase are other heme iron enzymes.

A second type of iron-containing compound consists of metabolically active compounds with an enzymatic function, but in which iron is not in the form of heme.[335] In mitochrondria, for example, nonheme compounds account for far more iron than do the cytochromes. A large portion of this iron is in a group of proteins designated as metalloflavoproteins. Metalloflavoproteins are involved in oxidative metabolism and include reduced nicotinamide adenine dinucleotide, succinate, and α-glycerophosphate dehydrogenases. Other enzymes of this group, such as monoamine oxidase, are not yet purified, but are presumed to contain iron.

The final type includes such enzymes as aconitase and microsomal lipid peroxidase, which do not contain iron yet require it as a cofactor. Iron in a loosely bound form is required for hydroxylation of proline and lysine in protocollagen, steps essential in the synthesis of collagen. This catalog is far from complete, and it is likely that additional compounds with important metabolic activities will be identified in the future.

Storage Iron

Iron within the body is stored as ferritin and hemosiderin, which together represent the second largest fraction of iron after hemoglobin. Ferritin is a specialized tissue protein designed for iron storage.[350] It has a molecular weight of about 450,000 and is composed of some 24 subunits. Within the central cavity of ferritin, masses of iron are deposited principally as ferric hydroxide, but they contain some phosphate as well. Hemosiderin is considered to be an aggregate form of ferritin and shows up as golden brown granules when seen by light microscopy. More detailed examination by the electron microscope suggests that these masses are composed of closely packed ferritin molecules. Because a variety of organic constituents are also included in the aggregate, it is postulated that these masses actually are disintegrated, ferritin-loaded, cellular organelles. These iron storage compounds are widely distributed in nature among different animals, plants, and fungi.[206, 398]

With a positive iron balance, the inflow of iron induces the synthesis of apoferritin, and, when it becomes loaded with iron, the ferritin molecule is protected against degradation as well.[245] Storage ferritin is synthesized by free polyribosomes, whereas serum ferritin may be produced by the endoplasmic reticulum of the reticuloendothelial cells and hepatocytes.[653] The ratio between ferritin and hemosiderin differs according to the total amount of iron stored within the cell. At lower concentrations of tissue iron, ferritin predominates; at higher concentrations, most of the iron found is hemosiderin. The exact mechanism whereby iron loads into the ferritin molecule or is released is largely unknown, although it may involve a $Fe(III) \rightarrow Fe(II) \rightarrow Fe(III)$ cycle. In mammalian species, diverse molecular forms of ferritin have been identified; in humans, different profiles are observed, according to whether the ferritin originates from heart, liver, or other tissues. The proposal has been made that these different species of ferritin represent various combinations of dissimilar subunits.[245]

IRON BALANCE IN HUMANS

Iron balance represents the relationship between iron absorbed and iron lost by the individual. The notion that iron exchange was very small originally came from careful chemical balance studies.[539] A better understanding was achieved through isotopic measurements of iron losses, more accurate quantitative estimates of iron requirements for growth and pregnancy, and measurements of iron absorption from food.

Physiologic Iron Losses

Losses in the male   Because of the small amounts of iron that men lose daily and the difficulties involved in distinguishing true losses from the presence of contaminating iron not actually excreted, it became evident that chemical measurements of excreted iron were unsatisfactory. Direct determination of radioactivity in urine, feces, and sweat after the intravenous injection of iron-59 permitted a distinction between excreted iron and contaminating fractions but presented considerable technical difficulties.[246] Such studies led to an estimate of about 1 mg/day of iron loss in the adult male, a figure that has been subsequently confirmed. Similar figures were later obtained by counting total body iron,[146, 636] but the relatively short half-life of iron-59 and the extremely limited daily loss of the isotope from the body made precise quantitation difficult. Long-term studies of body iron turnover employing iron-55, in which the specific activity of the circulating red cell mass was followed over several years, have provided the most accurate data.[281] A translation of these isotopic measurements into absolute amounts of iron lost, however, involved an assumption of complete body mixing and a definition of pool size. Analysis of the red cell activity curve showed an initial mixing component of about a 1-year duration and thereafter an exponential decrease at a rate of about 11%/year in the adult American male; this derivation was estimated to be equivalent to an iron loss of about 12 μg/kg/day or 1 mg for an 80 kg man.[323] Losses for other normal subjects in other countries studied by the same technique were about 14 μg/kg/day.[323]

The source of these basal iron losses has been categorized.[323] The largest single fraction, amounting to about 0.4 mg/day, consists of red cells entering the gut lumen. Additional losses of smaller magnitude, listed in Table 4-4, are derived from the iron content of the bile and exfoliated intestinal cells. These three sources of gastrointestinal iron are the basis for two-thirds to three-fourths of the total body loss. Urinary iron loss is of little consequence. Losses from the skin could not be accurately measured by chemical techniques because of surface contamination with iron. Radioactive measurements permit determination of the uptake by epithelial cells from transferrin, totaling about 2.5 μg/kg/day; this amount should represent the maximal iron available for loss through cell exfoliation, although some individual variation may occur. Perhaps the greatest fluctuations occur in blood loss from the intestinal tract, which may be increased by local irritants such as aspirin, alcohol, and other drugs. Other losses (skin, gastrointestinal mucosa, and urine) are affected by the level of serum iron, and therefore decrease during states of iron deficiency and increase in iron overload. However, the extent of iron losses probably is not reduced to less than 50% in iron deficiency and not elevated to more than twice normal in iron overload.

Special iron losses in females   Special requirements for iron face the adult female. Menstrual blood losses have been shown to average about 0.6 mg/day if distributed over the month.[180, 338] Therefore, the mean total iron loss in the menstruating female is about 20 μg/kg/day or 1.4 mg for the 65 kg female. This figure is consistent with radioactive measurements showing an overall loss of red-cell specific activity of 20%/year.[281] The distribution curve for menstrual blood loss is skewed: 11% of women lose over 80 ml blood/month, representing an iron loss of more than 1 mg/day.[338] This phenomenon is plotted in Figure 4-1. Thus, at least

Table 4-4.   Estimated major iron losses[a]

Source	80 kg male	
	μg/kg/day	mg/day
Gastrointestinal region		
Blood	5.0	0.4
Mucosa	1.0	0.1
Bile	2.5	0.2
Urine	1.0	0.1
Skin	2.5	0.2
Total	12.0	1.0

[a] Adapted from Green *et al.*[323]

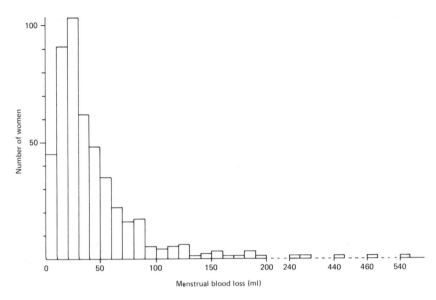

Figure 4-1. Distribution of menstrual blood loss in a population sample from women in Göteborg, Sweden. Reproduced from Hallberg *et al.*[338]

10% of menstruating women have iron requirements more than twice that of the male population.

The other unique loss for the female is associated with pregnancy. Two sets of requirements may be drawn up for pregnancy, summarized in Table 4-5. One represents iron needed during pregnancy, and the other, the actual iron lost with pregnancy and delivery.[50, 203] When distributed over 9 months, the requirement becomes 3.5 mg/day and the cost, 2.5 mg/day. The increased requirements are imposed largely during

Table 4-5.   Total iron requirements for pregnancy

	Average (mg)	Range (mg)
External iron loss	170	150–200
Expansion of red-blood cell mass	450	200–600
Fetal iron	270	200–370
Iron in placenta and cord	90	30–170
Blood loss at delivery	150	90–310
Total requirement[a]	980	580–1,340
Cost of pregnancy[b]	680	440–1,050

[a] Blood loss at delivery not included.

[b] Expansion of red cell mass not included.

the last 6 months of pregnancy, making the daily requirement during that period even greater.

Requirements during infancy and childhood   In infancy and childhood, attention is directed to the iron required for growth. The newborn is endowed with sufficient excess iron by the polycythemia present at birth and the hepatic iron stores to meet requirements for doubling body size during the first four months. In the premature infant, the reserve is greatly reduced—moreover, growth requirements are increased.[849] Through infancy, childhood, and adolescence, requirements of growth amount to 30 mg iron/kg increase in body weight. Information is limited concerning iron losses in infancy;[725] they are assumed to be at least equivalent to those of the adult on a kilogram basis. The frequent finding of guaiac-positive stools in infants and demonstrations of increased losses of radioiron in the stool suggest that infants lose more iron than do adults.[382] Much of this early bleeding may result from an immunologic reaction to cow's milk. A composite of average iron requirements throughout life is shown in Figure 4-2.[89]

Iron Intake and Absorption

Iron absorption is the product of the amount of iron in the diet, its availability, the influence of various luminal factors, and the behavior of

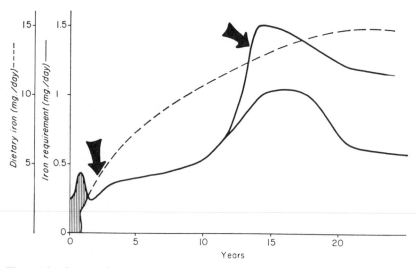

Figure 4-2.   Iron requirements in humans. The daily iron requirement through life is indicated by the continuous black line. For those over the age of 12, the line divides into the requirements of the female (upper line) and the male (lower line). The dotted line indicates available iron in the normal diet. The shaded area during the first year of life indicates a period of negative iron balance when the infant utilizes iron stores. The black arrows indicate the two critical periods when intake and loss are of similar magnitude. Reproduced from Bothwell and Finch.[89]

the intestinal mucosa. Much of the work on absorption has been carried out in animals, and although this research has been useful in developing certain general criteria, its relevance to humans is questionable because of species differences.

Mucosal regulation of absorption   That the intestinal mucosa regulates iron absorption has been known since 1945.[332] Most iron is absorbed in the duodenum and upper jejunum, although the mucosa throughout the intestine is capable of absorbing iron because of the favorable intraluminal conditions in that portion of the gut.

Oral administration of iron salts has permitted a more direct evaluation of the intestinal mucosa because highly available iron could be presented to the mucosal cell without the modifying effects of food. Body iron stores appear to be the most important influence on the absorptive capacity of the mucosa.[91, 194, 632] In subjects with marginally low iron stores whose hemoglobin or plasma iron concentrations were not changed, absorption was increased.[194, 456] Conversely, excess iron stores resulted in diminished absorption.

An increased role of erythropoiesis following hemorrhage, hemolysis, or exposure to decreased ambient oxygen tension also has been shown to increase iron absorption.[91, 185, 354, 544, 737] A return to more normal conditions will tend to decrease erythropoiesis and iron absorption.[91, 544, 833] Whereas it has been proved in animals that hemolytic anemia increases iron absorption, data for humans are less convincing. Only with disorders in hemoglobin synthesis, i.e., thalassemia and sideroblastic anemia, does iron absorption appreciably increase. Anemia *per se* has been described as increasing iron absorption.[545] Schiffer[691] found that markedly less iron was absorbed in patients with aregenerative anemia when their hemoglobin was restored by transfusion. Still another condition affecting absorption is inflammation, during which iron transport across the mucosa is reduced, accompanied by a similar decrease in iron release by hepatocytes and reticuloendothelial cells.

Although these factors and possibly others are known to modify iron absorption, the means by which it is accomplished is not known. Considerable effort has gone into the search for a humoral factor in the blood, yet convincing results have not been obtained.[781] Iron absorption may be regulated by iron received within the intestinal cells during their formation.[186, 187, 834] However, others have observed that the iron content of human mucosal cells is not appreciably changed during iron deficiency or overload.[16,43] Even if this suggestion is correct, the way in which the iron content of these cells is modified remains unexplained, inasmuch as plasma iron does not diminish with depleted iron stores, yet absorption increases.

The nature of the absorptive process is as yet unclear, although interesting observations have been made. Hübers *et al.*[392] found that iron-deficient rats have twice as much iron in their brush border as those of iron-adequate rats, suggesting that the brush border might be one point of control. Within the cell, several iron-binding substances have been found in the particle-free fraction of mucosal cell homogenates.[378, 621, 638, 864] A transferrin-like protein has been demonstrated that is found in greatest concentration in the upper part of the small intestine of rats. It is increased by iron deficiency and by pretreatment of iron with phenobarbital. The administration of phenobarbital also results in iron absorption, but it is not demonstrated in the intestinal mucosa of mice with sex-linked anemia who have impaired iron absorption.[393] This protein is not precipitated by antiferritin, antitransferrin, or antilactoferrin antisera. A similar protein has been described by Pollack.[638] However, proof adequate to establish any given protein as an important transport mechanism for iron has not yet been provided. Notable amounts of iron are held as ferritin within the mucosal cell in inverse relationship to the amount absorbed. This binding, however, is not thought to regulate iron absorption, but merely to hold iron not designed to enter the body, since the metal is lost when the mucosal cell is sloughed.

Luminal factors   The amount of dietary iron absorbed depends not only on the absorptive behavior of the intestinal mucosa, but on the effect of various luminal secretions and gastrointestinal motility. It has been demonstrated both in humans and in laboratory animals that absorption is usually reduced when gastric acid secretion is reduced or eliminated.[191, 527, 575] Similarly, the administration of hydrochloric acid to individuals with achlorhydria increases absorption in humans[191, 421] and animals.[843] Because inorganic iron compounds are more readily solubilized in acidic than in neutral or alkaline media, it is not surprising that the pH of the stomach is important to absorption. Gastric juices may contain other substances facilitating or inhibiting iron absorption, but on the basis of present data it is impossible to identify anything other than a nonspecific absorption of iron with mucoproteins secreted within the stomach.[564] Pancreatic secretions have also been thought to modify iron absorption, but the most recent animal and clinical studies do not support this idea. Thus it is inappropriate to assume any effect other than alkalinization of the pH in the upper bowel.[41, 236, 436, 574]

Alkalinization precipitates iron and makes it unavailable for absorption. At the same time, the digestion of food in the upper duodenum by pancreatic enzymes might well be expected to elaborate chelates such as cysteine and histidine, which might improve availability of iron for absorption. Studies of the physical form of iron in the stomach and duodenum indicate that iron is bound predominantly to macromolecules in the stomach, but micromolecular iron is found in the duodenum.[418] Ascorbic acid secreted into the stomach of some animals and into the duodenum from the bile has also been proposed as an aid to iron absorption, but the acid's quantitative importance is unknown.[183] The roles of the digestive process and of possible chelates secreted into the lumen are unclear.

Iron absorption also is affected by the motility of the gastrointestinal tract. Retarded gastrointestinal motility is accompanied by increased absorption, whereas operative procedures on the stomach that shorten emptying time are associated with decreased absorption.[147, 526] Such observations indicate that the "hopper" function of the stomach is essential for normal iron absorption.

IRON IN THE DIET

Humans ingest less iron than other animal species. Average dietary iron intake in developed countries is about 6 mg/1,000 cal; in developing countries, the content is more variable, averaging about 10 mg/1,000 cal.[50, 790] The higher values in developing countries may be a function of extrinsic iron that is contaminating food during its procurement and preparation. The iron content of various foods is listed in Table 4-6.

Employing previously discussed estimates of the amount of iron required to maintain balance, adult males in developed countries should be absorbing about 6% of dietary iron, and adult females should require about 12%. In developing countries, 4% and 8% absorption, respectively, would seem adequate. These requirements do not appear difficult to achieve when compared to absorption data obtained by the use of iron salts in fasting subjects; however, they are large when the limited availability of dietary iron is considered. In the past, the total iron content of the diet has been emphasized, and therefore the iron contents of individual foods. However, it is not practical to modify the iron content of a diet greatly by changing its composition except through fortification, and food iron content *per se* does not correlate well with the amount of iron absorbed.

Availability of Dietary Iron

Until recently, direct attempts to measure dietary iron availability have had limited success. Chemical balance studies have been technically difficult and imprecise. To understand the availability of food iron in humans, it was necessary to develop a methodology capable of overcoming certain problems. One complicating factor was the variation in iron absorption among individuals,[194] which included differences in iron balance affecting the mucosal setting for iron absorption, and meal-to-meal fluctuations, presumably related to physiologic differences in secretions and in motility of the gastrointestinal tract. Differences caused by variations in iron stores were controlled by carrying out comparisons in

Table 4-6.   Iron content of foods[a]

Food	mg iron/ 100 kcal	mg iron/100 g edible portion of food
Low Iron Content (<0.7 mg iron/100 kcal)		
Apples, raw	0.6	0.3
Avocado, raw	0.4	0.6
Beer	trace	trace
Bologna, frankfurters	0.6	1.8
Bread, white, unenriched	0.3	0.7
Butter	0	0
Cheese, cheddar	0.3	1.0
Cheese, cottage, creamed	0.3	0.3
Chocolate, semisweet	0.5	2.6
Codfish, broiled	0.6	1.0
Corn, fresh, cooked	0.7	0.6
Egg whites	0.2	0.1
Honey	0.2	0.5
Ice cream	trace	0.1
Lamb, loin chop, broiled	0.4	1.3
Margarine	0	0
Milk	trace	trace
Oil, salad or cooking	0	0
Orange juice	0.4	0.2
Peaches, canned in syrup	0.4	0.3
Peanut butter	0.3	2.0
Peanuts, roasted	0.4	2.2
Pears, raw	0.5	0.3
Potato chips	0.3	1.8
Rice, brown, cooked	0.4	0.5
Rice, white, unenriched, cooked	0.2	0.2
Salad dressing, mayonnaise	0.1	0.5
Salad dressing, mayonnaise-like	0.1	0.2
Sugar, white, granulated	<0.1	0.1
Sweet potatoes, baked	0.6	0.9
Wheat flour, white, unenriched	0.2	0.8
Medium Iron Content (0.7–1.9 mg iron/100 kcal)		
Almonds, dried	0.8	4.7
Apricots, raw	1.0	0.5
Beef, ground, cooked	1.1	3.2
Beef, round, broiled	1.9	3.7
Beef, T-bone steak, fried	0.6	2.7
Bread, white, enriched	0.9	2.5
Bread, whole wheat	0.9	2.3
Carrots, raw	1.7	0.7
Cereals, prepared breakfast, enriched at 2–7% USRDA per oz.	0.4–1.0	1.4–4.0
Chicken, dark, cooked	1.0	1.7
Chicken, white, cooked	0.8	1.3

Table 4-6.   (Continued)

Food	mg iron/ 100 kcal	mg iron/100 g edible portion of food
Cranberries, raw	1.1	0.5
Eggs, whole	1.4	2.3
Egg yolks	1.6	5.5
Gingerbread, made with enriched flour	0.7	2.3
Liverwurst	1.8	5.4
Molasses, light (first extraction)	1.7	4.3
Melon, cantaloupe	1.3	0.4
Oatmeal, cooked	1.1	0.6
Onions, mature, cooked	1.4	0.4
Oranges, raw	0.8	0.4
Peaches, raw	1.3	0.5
Pork, medium-fat, roasted	0.8	2.9
Potatoes, white, baked	0.8	0.7
Prunes, dehydrated, uncooked	1.3	4.4
Raisins, uncooked	1.2	3.5
Rice, white, enriched, cooked	0.8	0.9
Sardines, canned in oil, drained	1.4	2.9
Soup, canned, vegetable beef, ready to serve	0.9	0.3
Squash, winter, baked	1.3	0.8
Sugar, brown	0.9	3.4
Tuna, canned in oil, drained	1.0	1.9
Turnips, cooked	1.7	0.4
Walnuts, black	1.0	6.0
Wheat flour, white, enriched	0.8	2.9
Wheat flour, whole grain	1.0	3.3
High Iron Content (≥ 2.0 mg iron/100 kcal)		
Apricots, dried	2.1	5.5
Artichokes, cooked	4.2	1.1
Asparagus, cooked	3.0	0.6
Baby cereals, enriched	>13.5	>50.0
Beans, green, cooked	2.4	0.6
Beans, kidney, cooked	2.0	2.4
Broccoli, cooked	3.1	0.8
Caviar, granular	4.5	11.8
Cereals, prepared breakfast, enriched at 25% USRDA per oz.	3.8	15.0
Clams, hard shell	9.4	7.5
Coffee, instant, dried	4.3	5.6
Fish flour, from fish fillets	2.0	8.0
Fish flour, from whale fish	12.2	41.0
Giblets, chicken, fried	2.6	6.5
Heart, beef, braised	3.1	5.9
Kidneys, beef, braised	5.2	13.1
Lettuce, iceberg	3.8	0.5
Lettuce, romaine	7.8	1.4

Table 4-6.   (Continued)

Food	mg iron/ 100 kcal	mg iron/100 g edible portion of food
Liver, calves, fried	5.4	14.2
Molasses, blackstrap (third extraction)	7.6	16.1
Mung bean sprouted seeds, cooked	3.2	0.9
Oysters, fried	3.4	8.1
Peas, frozen, cooked	2.8	1.9
Soybeans, mature, cooked	2.1	2.7
Spinach, fresh, cooked	9.6	2.2
Squash, summer, cooked	2.9	0.4
Tomatoes	2.3	0.5
Turnip greens, cooked	5.3	1.0
Wheat bran, commercially milled	7.0	14.9
Wheat germ, commercially milled	2.6	9.4

[a] Courtesy of Elaine Monsen.

the same individual, employing two isotopes of iron. Sporadic variations were dealt with by studying about a dozen individuals of the same age and sex, or by administering repeated doses of isotope so as to determine absorption from the number of meals.[108]

Employing an adequate experimental design, the absorption of iron from single foods was evaluated. When compared to other forms of iron, hemoglobin was found to be particularly well absorbed by humans, and more important, was found to be uninfluenced by chelates that could block absorption of nonheme iron.[184] Heme is taken intact into the mucosal cell, where it is catabolized by heme oxygenase with the release of its iron.[840]

Many foodstuffs have been biosynthetically labeled and their absorption measured, as charted in Figure 4-3. Vegetal iron was poorly absorbed in normal subjects (1–10%), whereas meat iron was better absorbed (5–20%).[473] When these studies were extended to evaluate mixtures of tagged foods, it became apparent that individual foods interacted, affecting the availability of all the different forms of nonheme iron present. Indeed, virtually all nonheme iron in a single meal had the same availability. When added to a complex meal, a tracer dose of radioiron salt was absorbed in the same amounts as nonheme iron in individual food articles.[195] Similarly, a tracer dose of radioactive heme was found to be absorbed to the same degree as heme iron in food.[337, 474] These two extrinsic tags provided the first accurate means of determining iron absorption from the normal diet.

The two-pool, intrinsic tag method (heme and nonheme) has yielded estimates of food iron absorption that closely agree with estimates based on normal body iron losses. For example, meals composed of aliquots of all food consumed in a typical 6-week diet and doubly tagged with radioactive heme and nonheme iron were administered to 32 young men.[77] The total daily intake of iron in these men was 17.4 mg, of which only 1 mg was in the form of heme iron. Total absorption averaged 1.25 mg/day. Absorption of nonheme iron averaged 5.3% or represented 0.88 mg/day, whereas absorption of heme iron averaged 37% or 0.37 mg of iron. Similar data were obtained by Layrisse and Martínez-Torres,[474] who fed their subjects a meal of meat, black beans, maize, and rice containing a total of 4.5 mg iron. In their normal subjects, absorption from 1.5 mg of heme iron was 27% (0.34 mg), compared to an absorption of 3 mg of nonheme iron of 6% (0.12 mg). Total absorption from the meal was 0.46 mg iron. These studies underline the important contribution made by the heme iron to the diet.

If individual components affecting the availability of nonheme iron in a meal could be identified, absorption from any type of diet should be predictable. A number of inhibitors and enhancers of food iron absorption have been identified. Phytic acid is often considered a potent inhibitor of iron absorption; however, the effects of phytate on iron utilization are difficult to interpret because of several intervening factors. For example, the number of iron atoms sequestered by the phytate molecule has an effect. Iron from certain iron-phytate complexes is less available

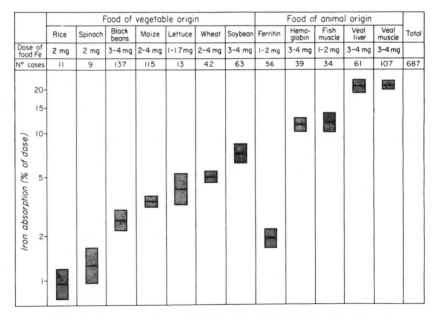

| | Food of vegetable origin | | | | | | | Food of animal origin | | | | | |
	Rice	Spinach	Black beans	Maize	Lettuce	Wheat	Soybean	Ferritin	Hemo-globin	Fish muscle	Veal liver	Veal muscle	Total
Dose of food Fe	2 mg	2 mg	3-4 mg	2-4 mg	1-1.7 mg	2-4 mg	3-4 mg	1-2 mg	3-4 mg	1-2 mg	3-4 mg	3-4 mg	
N° cases	11	9	137	115	13	42	63	56	39	34	61	107	687

Figure 4-3. Iron absorption from food biosynthetically labeled with radioactive iron. The horizontal line represents the geometric mean and the crosshatched area shows the limits of one standard error. Reproduced from Layrisse.[472]

for hemoglobin formation than that from more soluble iron salts.[198, 296, 298, 577] However, Morris and Ellis[567] recently reported that iron from monoferric phytate is readily available to rats. The research indicates that much of the iron in wheat is monoferric phytate. Moreover, wheat iron is readily available to rats[567] and humans.[198] Evidently, different iron-phytate complexes have differing availabilities, which must be taken into account when considering the effect of phytate on iron availability.

Another complication is that inorganic phosphates also depress iron utilization.[357] Thus, in experiments in which the total phosphorus content of the diet is not constant, it is not possible to separate the effects of phytate from those resulting from increased phosphate content of the diet.[204, 396] Phytate in vegetal food sources (such as legumes) is often complexed to protein or other components of the plant and may be less reactive than soluble phytate salts. The addition of soluble salts of phytic acid to diets has reduced iron availability, but naturally occurring phytates may not act the same way. For instance, Sharpe et al.[711] found that adding sodium phytate to milk reduced the iron absorption of human subjects. However, when the phytate was supplied by oats, no correlation was observed between the phytate content of the oats and iron availability. Similarly, Welch and Van Campen[836] found that the phytate concentration of soybeans was not correlated with the availability of the soybean iron to rats. Thus it is not at all certain if naturally occurring phytates have the same influence on iron utilization as soluble phytate salts.

Finally, recent reports of the effect of fiber on iron utilization have further complicated analysis. Several investigators have reported that substitution of whole-wheat bread for white bread or the addition of increasing quantities of wheat bran to the diet will depress iron utilization.[75, 293, 428, 711, 848] This phenomenon had been attributed to the phytate content of the whole meal or bran products. Yet phytin-free fiber can reduce iron utilization in humans[660] and rats.[738] Because wheat bran contains both fiber and phytate, it is hard to discern which is the critical factor in reducing iron utilization. At present it is difficult, if not impossible, to assess the individual effects of fiber, phytate, and phosphate on iron utilization.

Undoubtedly, a considerable number of substances may interfere with iron absorption. Calcium and phosphate salts and ethylene-diaminetetraacetic acid (EDTA), which are added to American diets as food preservatives, have been shown to reduce iron availability.[200, 557] At present, two food components have been shown to enhance absorption. The availability of nonheme iron increases when the meal contains animal tissues.[478, 532] For example, the substitution of 100 g beef for an equivalent amount of egg albumin in a test meal increased absorption more than five times.[199] Thus meat is important not only as a source of heme iron, but as an enhancer of absorption of nonheme iron. Muscle from animals and fish and liver have this property, whereas milk and cheese do not. Ascorbic acid is another potent enhancer of iron absorption because it can reduce iron *and* form a chelate with ferric iron at low pH, effects that maintain solubility at the higher pH of the duodenum. Recent studies have shown an enhancing effect on nonheme iron absorption by a relatively small amount of ascorbic acid that was either contained in food or added to food during its preparation.[687] For example, 60 mg ascorbic acid added to a meal of rice more than tripled absorption of iron,[686] and 150 g of papaya containing 66 mg ascorbic acid increased iron absorption more than fivefold when taken with a meal of maize.[476] Such studies of dietary iron availability employing the extrinsic tag have clarified information on iron absorption and loss and pointed out that availability may be even more important than content in determining the amount of iron absorbed from food.

Some limitations in the extrinsic tag technique as applied to nonheme iron absorption have become evident. These shortcomings do not invalidate its use as described, but need to be considered in special situations. Certain iron salts used in fortification, such as pyro- and orthophosphates are incompletely miscible with the nonheme dietary iron pool.[198] Certain forms of food iron, such as ferritin and hemosiderin with their masses of ferric hydroxide, are also incompletely miscible and it is presumed that soil iron would be even more so.[477]

Balance and Stores

From the information available on iron requirements and available iron in the diet, humans appear to be uniquely restricted in external iron exchange. Basal exchange in the adult male of 12 μg/kg/day may diminish to as little as 6 μg/kg/day with iron deficiency, and the adult menstruating female, with an average requirement of about 20 μg/kg/day, may decrease her daily requirement to about 15 μg/kg. An iron-deficient American male should be able to absorb a maximum of about 50 μg of iron/kg/day from his average diet, whereas the iron-deficient menstruating female might be expected to absorb about 35 μg/kg/day. These availability estimates for either sex may be halved if the diet does not contain enough meat or ascorbic acid. It is apparent that the male population has a considerable margin of safety, whereas a major portion of the female population must be at risk. A prevalence of a 4% iron deficiency in the adult male population (most of which is related to pathologic bleeding), compared to a 20% iron deficiency in the adult female population, reflects this distinction.[192] The inability of most pregnant women to meet iron requirements is acknowledged by the routine therapeutic administration of iron salts. A high prevalence of iron deficiency is also found in infancy, in which requirements of growth outstrip available dietary iron.[203] There is no question that iron balance in menstruating and pregnant women and in infants is perilous.

One of the best indicators of an individual's iron balance is the extent of ferritin and hemosiderin stores. They are normally found equally divided among the hepatocytes, reticuloendothelial cells, and striated muscle.[778] These stores have been evaluated in various ways. The most direct but least practical method is phlebotomy, which mobilizes iron stores; approximately 1 g of storage iron has been measured in the adult male and one-third that much in the female.[43, 611, 651] Another approximation of storage iron is provided by examining a marrow aspirate for hemosiderin.[734] In healthy individuals, a negative correlation has been shown between the hemosiderin of the marrow and the total iron-binding capacity of the plasma.[830] However, the effect of other factors, particularly protein malnutrition, on the iron-binding capacity, vitiate the usefulness of this measurement in evaluating stores. One of the more informative procedures has been the assay of a postmortem liver specimen for nonheme iron. The results of some 4,000 determinations in 18 different countries have been reported by Charlton et al.,[167] and mean values for their male subjects are summarized in Table 4-7. Assuming that the liver iron of such individuals represented about one-third of storage iron, estimates of the size of body iron stores could be made, as well as comparisons among different populations.

Table 4-7.   Hepatic storage iron concentration[a]

Countries	Males		Females	
	No. of subjects	Mean (mg/kg)	No. of subjects	Mean (mg/kg)
Great Britain	182	158	128	156
Sweden	422	159	300	129
Czechoslovakia	187	212	187	182
United States	232	188	121	144
South Africa				
Caucasian	97	258	42	161
Bantu	318	818	133	270
Rhodesia				
Bantu	142	681	76	228
Nigeria				
Bantu	87	189	96	187
India	203	112	67	79
New Guinea	110	106		
Venezuela	204	203		
Brazil	74	187		
Mexico	151	198		

[a] Adapted from Charlton et al.[167]

Immunologic methods have been developed to measure serum ferritin.[419, 549] The values obtained in several of the experimental situations showed a close relationship to iron stores. That is, the geometric mean of about 90 in males as compared to 30 in females was proportional to the difference between iron stores in the two sexes. A relationship has been shown between serum ferritin and iron absorption[196] and between serum ferritin and iron stores as determined by phlebotomy.[815] Only in hepatic disease and inflammation does the ferritin level appear to deviate from these relationships.[495]

INTERNAL IRON METABOLISM

Internal iron exchange in mammals is dominated by iron requirements for hemoglobin synthesis. The iron content of the erythron is about 1 g/kg tissue, some two orders of magnitude greater than the essential iron content of other body tissues. In the design of internal iron exchange, therefore, some means had to be created whereby the disparate needs of individual tissues could be met and excess iron could be stored (accomplished through the function of a plasma protein, transferrin).[562] The phylogenic appearance of this plasma protein coincided with the appearance of hemoglobin in red cells. When, as a genetic oddity, transferrin is absent in a human, the unique finding of iron deficiency anemia despite iron overload of other body tissues is observed. This phenomenon is clear evidence of the protein's essential function in iron exchange.[320]

The first steps in binding appear to be the interaction of transferrin with iron, and the release of three protons; the second step involves the inclusion of bicarbonate in the iron transferrin complex and is associated with the development of a salmon-pink color. The reverse reaction occurs with iron release. Iron binding above pH 7.2 is maximal, with a binding constant in plasma of approximately 10^{24} M^{-1}.[9] As the pH is reduced to <6.5, iron starts to dissociate, and it is nearly completed at pH 4.5.[749] *In vivo* transferrin shuttles iron back and forth between body tissues without being used up itself. The behavior of the two iron-binding sites has been the subject of some controversy: it is undecided if they participate independently and equally in the binding and release of iron, or if differences in affinity exist between them and certain binding tissues.[8] Recent evidence favors the homogeneous behavior of transferrin iron, with specific tissue receptors decisively determining the exchange.[642] The ability of transferrin to bind a variety of trace metals is noteworthy;[7, 759, 861] however, the patterns in which these metals are released are quite different from that of iron and they are not assimilated by immature erythrocytes in any appreciable amount.

Current methods for measuring plasma iron largely exclude hemoglobin and are therefore assumed to represent isolated transferrin iron.[402]

However, they will measure part of parenterally injected preparations, such as iron dextran. The normal level of human plasma iron is about 100 μg/dl and saturation of transferrin with iron is about 33%.[89] Whereas transferrin concentration is relatively stable from hour to hour and day to day, serum iron in the normal individual may vary over 24 hr between 50 and 200 μg/dl. Diurnal variations are normal, with morning values at least 50% higher than evening values. The higher levels of plasma iron commonly found in animals are related to a much greater iron absorption, and fasting will reduce the level to about 100 μg/dl.[193] Decreases in plasma iron are associated with exhaustion of iron stores, inflammation, increased erythropoiesis, and reduction in transferrin concentrations.[89] Elevations in plasma iron are caused by increased absorption, hepatic damage, increased blood destruction (in particular, ineffective erythropoiesis), and the parenteral administration of iron. Figure 4-4 sets forth relationships between plasma iron and transferrin concentration observed in various diseases.

The erythron contains 90% of essential body iron in most mammalian systems and is therefore the primary user of transferrin iron. Iron is taken up by the immature red cell and incorporated into its hemoglobin, and remains within the cell for its life span. Uptake initially involves the adherence of the transferrin iron complex to membrane receptors. Then the complex is internalized in microtubules and the iron is removed, followed by the return of transferrin to the surrounding plasma.[363, 434] Within the erythroid cell, more than 80% of the iron is

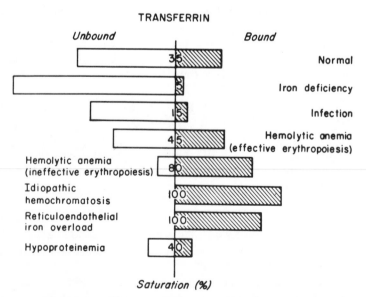

Figure 4-4. Relationships between transferrin and plasma iron in disease. Reproduced from Bothwell and Finch.[89]

delivered to the mitochondria, where it is converted into hemoglobin.[600] Excess iron is deposited in the cytoplasm in the form of ferritin aggregates. Although a coordination exists between heme and globin synthesis that keeps their production balanced, the ratio of production of these moieties has little effect on the uptake of iron by the developing red cell. If hemoglobin synthesis is reduced, excess iron will accumulate in ferritin stores with aberrant globin synthesis or in mitochondria with abnormal heme synthesis.

Usually, transferrin iron above 50 μg/dl provides the erythroid marrow with the amount of iron it needs for hemoglobin synthesis. However, the adequacy of iron supply is influenced by changes in requirement by the erythroid marrow as well as by fluctuations in plasma iron concentration. As production increases, the minimal level of plasma iron required to support erythropoiesis increases proportionately.[422] The adequacy of iron supply versus erythron needs may be monitored by red cell protoporphyrin, which increases as the ratio of plasma iron supply to erythron needs falls.[462]

Quantitative aspects of iron exchange between transferrin and the erythron have been studied in detail with iron isotopes.[283] The basal turnover of this fraction depends on the life-span of the red cell and varies from about 3%/day in mice to 0.8%/day in humans. However, with the wastage iron of erythropoiesis, iron turnover within the erythron is greater than that calculated from red cell life-span. Each day in the adult human, some 25 mg of iron is used by the erythron.[283] The anatomic location of the erythroid marrow may be visualized by the positron camera after injection of iron-52,[800] and the uptake and release of iron may be measured by surface probes placed over marrow-rich areas, such as the sacrum.[284]

The other component of the erythron iron circuit is the reticuloendothelial cell, which processes the iron of senescent or defective red cells and returns it to plasma transferrin. The wastage iron of erythropoiesis is released to the reticuloendothelial cell at the time of red cell maturation. Within the reticuloendothelial cell, red cell hemoglobin is catabolized. Its iron is either directed to the cell membrane to be taken up by transferrin, an exchange in which ceruloplasmin participates,[654] or it is deposited within the cell as ferritin. In dogs, about one-half of catabolized red cell iron goes by each pathway;[279] in humans, only one-third is held in storage.[280] The partition between these two pathways is inconstant, and the variation is thought to be largely responsible for the diurnal variation in plasma iron. During inflammation, the greater deposition of ferritin iron within the reticuloendothelial cell results in lowered iron concentrations in plasma.[279, 368]

An alternate pathway for transferrin iron uptake involves the liver and other parenchymal tissues.[283] The hepatocyte accepts and stores iron when uptake by the erythron decreases, as illustrated by *in vivo* profiles obtained in patients with aplastic anemia.[283] Studies in the rat show some localization in muscles, skin, and subcutaneous tissues, but the largest amount resides in the intestinal mucosa, reflecting the ability of that species to excrete iron through the gut.[169] The only other well-delineated pathway in humans is the exchange of plasma iron with the extravascular transferrin pool.[563] The total format of internal iron exchange is shown in Figure 4-5. The diagram does not illustrate the uptake by hepatocytes of the hemoglobin iron released through intravascular hemolysis of red cells.[367]

Isotopic studies of plasma iron kinetics have provided a better appreciation of these various pathways and their functional interrelationships.[156] More detailed data on kinetics are available for humans than for other mammals. Iron turnover through the plasma is considerably greater than was expected for the needs of the circulating red cell mass, an excess largely explainable by a reflux of radioiron from tissues.[609] Part of this reflux represents the wastage iron of erythropoiesis, amounting to about 24% of plasma iron turnover in humans with a half-time of about 7 days.[197] The second portion of this reflux, 8% of plasma iron turnover with a half-time of about 7 hr, relates to extravascular exchange of transferrin iron. In humans, absorbed iron comprises only about 3% of plasma iron turnover and has little effect on internal exchange. By contrast, as much as 50% of plasma iron turnover in the rat may be derived from absorption.[193] Detailed ferrokinetic analyses of humans have shown that the nonerythron turnover fraction, including extravascular flux and parenchymal uptake, is proportional to the amount of plasma iron.[197] Thus, by an empirical calculation of this fraction and its subtraction from the plasma iron turnover, the amount of erythron iron turnover can be ascertained accurately. The erythron in a variety of disorders has shown a large range of iron uptake: from 0 to 4 mg/kg (normal, 0.4 mg/kg).[283] Parenchymal uptake is more limited and reaches maximal

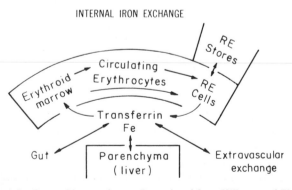

INTERNAL IRON EXCHANGE

Figure 4-5.   Internal iron exchange. Reproduced from Hillman and Finch.[380]

values of about 0.6 mg/kg, and only when plasma iron is high and transferrin is saturated. It is not possible to measure iron stores by short-term
kinetic measurements, because the early mixing of radioiron with stores
is limited. As erythron iron cycles through the reticuloendothelial cell, it
will mix with ferritin, but approximately a year is required for stores to
reach equilibrium.[323] Hemosiderin stores may not exchange iron. Kinetic
measurements have revealed the difference between reticuloendothelial
and parenchymal iron storage.[283] Reticuloendothelial iron stores are the
products of red cell catabolism, whereas parenchymal iron stores are
created when elevated plasma iron leads to an increased hepatocyte
uptake. Hepatocyte loading may be further augmented in hemolytic
states when the hemoglobin and ferritin are transported to them.

IRON BALANCE IN ANIMALS

Information on iron balance in mammals is more fragmentary than data
for humans, but marked differences are known to exist that must be
taken into account when interspecies inferences are drawn. For example,
the rat has an active excretory mechanism whereby plasma iron is taken
up by the gut mucosa and lost from the body when mucosal cells are
exfoliated.[169, 186] Under basal conditions, 10–15% of the plasma iron
turnover follows this pathway, and the amount may be increased when an
animal is given iron supplements. A rat's dietary iron intake is greater
than 100 times that of a human's when expressed on a per kilogram
basis.[193] Absorption of nonheme iron is more efficient that in humans,
whereas heme iron is less well absorbed. As much as 50% of the iron
entering a growing rat's plasma may be derived from absorption, and the
plasma iron is elevated to two to three times that of humans as a consequence. Growth requirements are much greater, and erythropoiesis in the
newborn rat is two to four times greater than that of the fully grown
animal when expressed per kilogram body weight because of the requirements imposed by expansion of blood volume. These differences make
comparison between rats and humans difficult. The use of the rat to
evaluate iron availability of foods is equally suspect. From the standpoint
of iron balance, rats are far better off than humans. Not only is iron deficiency virtually nonexistent, but it has been impossible to produce a pattern of tissue damage similar to that seen in humans with iron overload.

These differences observed in the rat apply to a greater or lesser
degree to other animal species.[787] Differences are to be expected in the
adequacy of iron balance at different ages, the rate of iron exchange, and
the ability to absorb heme versus nonheme iron. Some animals, such as
the pig, are born with very low iron stores and others, such as the rabbit,
with very adequate reserves. Animals generally seem to have little

problem in maintaining iron balance unless removed from their natural environment and placed on an artificial diet.

Although information concerning iron metabolism of most farm animals is limited, enough concern has arisen about their nutritional needs to define dietary iron requirements. Swine appear to have the greatest problem in maintaining iron balance, due to their very rapid growth and low iron stores at birth.[787] Estimates of food iron required for baby pigs range from 60 to 125 mg iron/kg dry diet. Matrone et al.[536] reported that shoats fed a milk substitute needed 125 mg/kg of food iron to maintain normal growth and hemoglobin. Braude et al.[101] stated their estimates in another way: baby pigs must retain 21 mg of iron/kg live weight increase to maintain normal iron status. All these estimates of iron requirements are complicated by the variable utilization of dietary iron. Matrone et al.[536] found that 30% of the iron in swine diets was utilized for hemoglobin synthesis, whereas Ullrey et al.[785] found that 82%, 61%, and 50% of the food iron was utilized when the diets contained 25, 35, and 125 mg iron/kg diet, respectively. Braude et al.[101] reported from 30% to 60% retention of orally administered iron, according to the method of dosing. As a result of these other studies, the National Academy of Sciences has recommended that diets for growing pigs contain at least 80 mg iron/kg dry matter.[584] Requirements for adult swine are not known; however, in herds in which parasitism is not a problem, natural feed ingredients should provide adequate iron intake.

The iron requirements of poultry for growth have been studied. Hill and Matrone[379] suggested on the basis of blood data that a diet containing 50 mg iron/kg would meet the chick's requirement. Subsequently, Davis et al.[226] reported that 4-week-old chicks required between 65 and 105 mg iron/kg diet to maintain normal weight gains and hemoglobin levels. They later observed that chicks fed a soybean protein or a casein-gelatin based diet required 75–80 mg iron/kg of diet.[227] Thus, it appears that diets for growing chickens should contain a minimum of 50 mg iron/kg and that 80–100 mg of the element are required to provide an adequate safety margin. Reliable estimates for mature chickens are not available; however, the demand for iron by laying hens is very high. An average hen's egg contains about 1 mg of iron—thus laying hens require some 200 mg iron/year for egg production. Although hemoglobin levels will decline slightly with the onset of laying, no striking reduction of body iron stores has been measured in laying hens.[656] Because iron deficiency is not considered a practical problem in poultry, it should be safe to assume that diets composed of natural feedstuffs supply enough iron for growing and laying chickens.

Veal calves fed unsupplemented whole-milk diets rapidly developed anemia. Matrone et al.[535] fed calves a diet of whole milk that provided 30 or 60 mg iron/day depending on whether or not iron supplements were added. Both diets were adequate in maintaining normal growth and hemoglobin levels from birth to 40 weeks of age. In this case, 30 mg iron/day was sufficient for calves in the age group studied; however, the dose is considerably smaller than the value of 100 mg/day suggested by Blaxter et al.[79] Bremner and Dalgarno[102] reported that for calves on a fat-supplemented skim-milk diet, 40 μg iron/g diet prevented all but a very mild anemia when fed from 17 days to 11 weeks of age. The hemoglobin levels of these calves were increasing at the end of the experiment, and the authors speculated that they would have been normal had the dietary regime been continued to the calves' normal slaughter weights. They found that similar results were obtained when iron was supplied as ferrous sulfate, ferric citrate, or ferric EDTA; however, iron phytate was not utilized as well as the more soluble iron compounds. Requirements for mature cattle are not known; however, it has been estimated that milking cows require 50–60 mg iron/day and that pregnant cows need 60–80 mg/day.[453] These levels should be met by virtually any diet normally consumed by these animals.

Data on iron requirements of sheep are limited. Lawlor et al.[471] experimented with sheep fed purified diets containing levels of iron ranging from 10 to 280 ppm. They found that 10 ppm was clearly inadequate for growing lambs, 25 ppm did not support normal growth, and 40 ppm appeared to be adequate. They concluded that the minimal requirement for iron in growing lambs was in excess of 25 ppm but not more than 40 ppm. No experimentally derived estimates of iron requirements for mature sheep are available; however, mature sheep need about 10–15 mg of iron/day.[453] This requirement is met when the animal's feed averages 10–15 ppm iron. As forages and feed grains uniformly contain considerably more than 15 ppm iron, healthy sheep fed normal diets should have no trouble in meeting their iron requirements.

These are several extensive compilations of feed analyses that include data on iron content.[550, 582-584] Underwood[786] summarized the iron content of common animal feeds: grasses, 100–250 mg/kg; legumes, 200–300 mg/kg; cereal grains, 30–60 mg/kg; and oilseed meals, 100–200 mg/kg. Except for milk and milk by-products, feed by-products of animal origin contain high levels of iron. The following figures are representative of the iron content of some of these materials: bloodmeal, 3,108 mg/kg; fishmeal, 381 mg/kg (range, 210–800); meatmeal, 439 mg/kg (range, 180–920); dried skim milk, 52 mg/kg (range, 5–104) dry material. Additionally, many inorganic materials fed to animals as sources of calcium and phosphorus—i.e., ground limestone, rock phosphate, ground oyster shells, and dicalcium phosphate—contain high levels of iron, frequently more than 500 mg/kg.[786] The high levels of iron in common animal feeds explain why iron deficiency is not generally a practical problem in farm animals, except in young animals consuming only milk or milk-product diets. Specific information, however, on the availability of these different iron salts is lacking.

5

Iron Deficiency

PREVALENCE

Worldwide and in the United States, iron deficiency is the most common nutritional problem and cause of anemia. However, estimates of prevalence are difficult to determine with precision because of problems in defining what constitutes iron deficiency.

Three stages of iron deficiency can be defined. The mildest form, depletion of iron stores (sometimes termed pre-latent iron deficiency), is characterized by decreased or absent storage iron. No other abnormalities are identifiable. To detect pre-latent iron deficiency, iron stores must be measured in one of three ways: by phlebotomy to the point of iron-deficient erythropoiesis,[353, 815] which yields a quantitative assessment of iron stores (the amount of iron in the hemoglobin removed); by a semiquantitative estimate derived from the amount of stainable hemosiderin in bone marrow aspirates or biopsies;[657] or by measuring an increase in intestinal iron absorption, which is correlated with decreasing iron stores.[358, 863] Because none of these methods has been practical for large population studies, considerable interest has arisen in the recently introduced serum ferritin measurements. These sensitive immunoradiometric assays appear to reflect iron stores.[715, 746, 789, 790, 815] Although few reports have been published to date on the efficacy of serum ferritin concentrations in detecting iron deficiency in large population studies,[140, 192, 862] the approach seems a promising one.

The second stage, latent iron deficiency, ensues after iron stores are depleted. It is characterized by reduced serum iron concentration, elevated transferrin concentration, and a resultant drop in transferrin iron saturation to less than 15%.[35] Concomitantly, hemoglobin synthesis is impaired by iron lack, causing erythrocyte protoporphyrin concentration to rise to more than 70 μg/dl.[462] However, in this stage of iron deficiency, circulating hemoglobin concentrations remain within the normal range, although treatment with iron results in slight hemoglobin increases.[305]

Overt iron deficiency anemia, the third stage, is triggered when the restricted hemoglobin synthesis contributes to a measurable decrease in the concentration of circulating hemoglobin or in the volume of packed erythrocytes (hematocrit). Tests of large populations have been able to discern only overt or latent iron deficiency. Studies that add serum ferritin measurements to transferrin saturations and erythrocyte protoporphyrin determinations may detect the earliest stage of iron deficiency, which will provide increased precision in defining iron deficiency in population surveys.[140, 192, 419, 420, 495]

Another difficulty in estimating the prevalence of iron deficiency in population studies is the uncertainty about what constitutes an exact range of normal hematologic values. Anemia is difficult to define with precision for different age groups. Most "normal" values are arbitrary and represent the best judgments of committees of experts; indeed, sharp demarcation of continuous variables may be impossible.[50] Another problem stems from sampling biases: populations with varying susceptibilities to iron deficiency must be included to obtain a valid distribution rather than projecting population data from easily sampled groups. Older studies often suffered from the loose attribution of iron deficiency as the cause for most of the anemias detected. Multifactorial causes of anemia and the difficulty of separating serum iron changes associated with infection or chronic disease from those of iron deficiency contribute to the problems of interpreting prevalence data. These complexities are discussed by Beaton[50] in a thoughtful review of the epidemiology of iron deficiency.

When values of transferrin iron saturation of less than 15% are used as a criterion of iron deficiency,[35] the prevalence of iron deficiency in selected populations of various countries is shown in Table 5-1.[863] These results—from relatively small numbers of subjects in each group (48–220)—emphasize a population particularly at risk (pregnant women), but they also show the incidence of iron deficiency in men and nonpregnant women. The study also demonstrates the discrepancy between iron deficiency as defined by low transferrin saturation, and anemia as defined by hemoglobin levels below arbitrary standards. In addition, other or coincidental causes of anemia, such as those based on deficiencies of vitamin B_{12} or folic acid, are identified. By the criterion of low transferrin saturation, 40–99% of the pregnant women sampled from the six countries studied were iron-deficient. Nonpregnant women showed 11.4–42.5% iron deficiency, and men showed a range of 0–8.8% iron deficiency.

Other large population studies have confirmed the prevalence of iron deficiency in many countries of the world.[189, 862] In the United States, information about the prevalence of iron deficiency has been derived largely from patients seen by physicians,[203] but two large populations have been surveyed.[789, 790] The first was designed to obtain information about nutritional deficiencies in a low-income population of the United States.[790] The second study, the "HANES" preliminary report, provides a more balanced appraisal of nutrition in the United States.[789] Again, discrepancies are apparent between different criteria of anemia (such as hemoglobin and hematocrit values) and raise questions concerning the standardization of methods, the criteria of iron deficiency, and differences among populations studied. Iron deficiency, estimated from a serum transferrin saturation of less than 15%, was observed in about 10% of children aged 1–5 years.[789] Projecting this percentage to the total population of children in the United States, the authors estimated that a total of 383,000 children met this criterion of iron deficiency. Adolescent girls and women during their reproductive years were far more likely to be iron deficient than men, consistent with physiologic iron losses of menses and pregnancy (see Chapter 4). Considerable additional information is included from separate analyses of subgroups of different races and economic status. However, some information of interest to epidemiologists is hidden by analyses that define low transferrin saturation differently ($<15\%$ and $<20\%$) for different groups divided by age and sex. The result of these arbitrary settings tends to minimize the prevalence of iron deficiency in the various categories. Presumably, the defects of the HANES study can be rectified and finer analyses of the effects of age, sex, and pregnancy will eventually be made. If the defects of the preliminary report are corrected, this study should be a milestone in the collection of data on the prevalence of iron deficiency in the United States.

Another means of estimating the prevalence of iron deficiency in population samples is the measurement of hepatic iron stores. When iron deficiency is defined in terms of nonheme iron concentration of <50 $\mu g/g$,[746] measurements in a group of adults dying accidentally indicate that 24% of women and 8% of men in the United States have iron deficiency. Comparable figures for a Canadian population are 20% and 5%, respectively.[709] Many thousand samples of liver for iron determination have provided evidence of the widespread prevalence of iron deficiency in other countries.[29, 162, 167, 611]

By almost any criterion, iron-deficiency anemia and, to an even greater extent, latent iron deficiency are worldwide problems affecting millions of people.

CAUSES OF IRON DEFICIENCY

Most cases of iron deficiency in humans are caused by insufficient dietary iron, iron losses, or malabsorption of iron. Although each of these causes has important health implications, the problem of insufficient iron assimilation is probably the one most amenable to solution.

Insufficient Dietary Iron

People fail to assimilate enough iron from the food that they eat because of absence of iron in the food itself; insufficient quantities of iron-containing foods; ingestion of cleaner food to which no extraneous iron has been added during processing; interference with absorption due to

Table 5-1. Prevalence of anemia[a] and of iron, vitamin B_{12}, and folate deficiencies[b]

Country	Subjects	Proportion of cases with anemia (%)	Proportion (%) of cases with the following deficiencies			
			Serum iron (<50 µg/dl)	Transferrin saturation (<15%)	Serum vitamin B_{12} (<80 pg/ml)	Serum folate (<3 ng/ml)
Israel	Women pregnant	47.0	36.0	46.3	2.0	6.3
	nonpregnant	29.0	15.0	11.4	0	5.1
	Men	13.6	12.1	8.8	0	1.6
Poland	Women, pregnant	21.8	31.4	40.0	0.5	1.4
India (Delhi)	Women pregnant	80.0	10.0	51.7	49.0	
	nonpregnant	64.3	9.5	25.8	26.7	
India (Vellore)	Women pregnant	56.0	99.0	99.0	0	9.0
	nonpregnant	35.0	51.0	42.5	3.0	0
	Men	6.0	7.1	5.1	0	2.0

Mexico	Women	pregnant	26.6	30.5	61.2	7.1	6.5
		nonpregnant	11.7	22.1	28.1	0	6.0
	Men		0.9	6.2	3.6	1.1	3.5
Venezuela	Women	pregnant	37.0	56.8	59.7	23.0	15.1
		nonpregnant	14.9	14.0	18.9	1.0	9.5
	Men		1.9	7.5	0	2.0	18.8

[a] The following hemoglobin concentration per deciliter of blood are considered to indicate the presence of anemia:

Children, 6 months-6 years	<11 g/dl
Children, 6–14 years	<12 g/dl
Adult males	<13 g/dl
Adult females, nonpregnant	<12 g/dl
Adult females, pregnant	<11 g/dl

[b] Derived from the World Health Organization.[863]

natural or artificial inhibitors of dietary iron absorption; and reduced intake of substances that promote iron absorption. Often a combination of these factors will play a role in reducing the assimilation of iron from food (see Chapter 4).

As can be seen from Table 4-6, many foods are poor in iron. Prominent in this list are milk and milk products, many fruits and vegetables, sugar, and salad and cooking oils. There is little wonder that infants and small children, whose major caloric intake is from cow's milk, are at risk of iron deficiency. Intake of mother's milk instead of cow's milk improves the absorption of iron. Selecting foods low in iron may be a function of habit, economic necessity, or religious and cultural reasons, but the result is the same: an iron intake insufficient to balance losses and eventual development of iron deficiency.

The quantity of food consumed by most people is based on their caloric requirements.[50] Accordingly, men and boys eat more food than women and girls. Although iron content and availability vary considerably in individual foods, the mixtures of foods people eat provide a fairly constant iron concentration in relation to caloric intake. Beaton[50] has tabulated this iron:kilocalorie ratio from calculated and measured patterns of dietary iron consumption in many countries. In the United States, the iron concentration in milligrams per kilocalorie ranged from 4.9 to 6.3 in one study[50] and was about 6.0 in another.[789] Iron intake calculated in relation to energy requirements for men expending 3,000 kcal yields about 18 mg dietary iron from the average American diet. Women whose energy expenditure is approximately 2,000 kcal would be expected to consume about 12 mg dietary iron. This latter value is close to that actually measured in a group of 97 Canadian women[51] and exceeded the mean value (9.9 mg) recorded for 13 young American women.[558] Caloric intakes of children are related to age, but vary within the range of 1,500–2,000 kcal,[789] which would provide 9–12 mg iron from diets similar to those consumed by adult members of a family. Thus, lower caloric intakes in women and children place them at risk of insufficient dietary ingestion of iron.

The possibility that iron intake is somewhat dependent on the addition of extraneous iron during food preparation was emphasized by the major discrepancies found in calculated and chemically analyzed iron concentrations of diets sampled primarily from institutional dietary kitchens in a variety of foreign countries.[50] Contamination of acid foods from iron cooking pots has been demonstrated by Moore[559] and the addition of iron to Kaffir beer consumed by the Bantu is well documented.[83] The converse, reduced iron contamination of foods from the use of stainless steel and glass cooking vessels, seems to be more widespread in developed countries where modern food processing prevents marked contamination. Several studies have shown close correlation when iron content of diets from actual chemical measurements[51, 558, 865] is compared to calculated iron content from food tables.[824]

Availability of iron in foods depends considerably on absorption from single food sources as well as the interaction of foods on the absorption potential of the iron compounds they contain. The graph of foods labeled with radioactive iron shown in Figure 4-3 indicates a wide range of variation in absorption of food iron.[472] As a rule, iron in cereals and vegetables is poorly absorbed, whereas iron from animal sources, especially heme-containing compounds, is absorbed much better. Furthermore, absorption of vegetable food iron is almost doubled when it is ingested with heme from meat, liver, or fish[478, 532, 534] or when ascorbic acid is added.[77, 145, 456, 687] Iron absorption reflects that of the least well absorbed vegetable when two vegetables are mixed, and is the same as the vegetable iron when up to 60 mg of a ferric or ferrous iron salt is mixed with a vegetable.[78, 195, 474, 475] Addition of a chelating agent (desferrioxamine) diminishes the absorption of vegetable iron.[195, 456] Heme iron compounds from meat, liver, and fish are not affected by desferrioxamine or ascorbic acid.[195, 474]

These results have been interpreted as evidence of two major iron pools, i.e., one a heme and one a nonheme pool, with the various iron compounds in each pool having the same percentage of absorption. Each pool is affected by substances present in food that inhibit or enhance the metal's absorption. Furthermore, iron compounds used for food fortification are absorbed to the same extent as foods in the nonheme pool.[195, 475]

The distinction of dietary iron according to iron pools has permitted measurements of iron absorption from complex dietary mixtures in which the interactions of iron from many sources were once difficult to discern. A tracer dose of iron salt, e.g., iron-59 chloride, to label the nonheme pool mixed with iron-55-labeled hemoglobin provides separate measurements of the nonheme and heme pools of a dietary mixture and an accurate estimate of percentage and total absorption from the two forms of iron. This extrinsic method of labeling also makes it possible to measure absorption from iron compounds used for food fortification as well as from the naturally occurring iron of foods. Diets with an adequate total amount of iron but containing little heme, which enhances absorption of the rest of the iron intake, may not be adequate to meet nutritional requirements. Thus, fortification of foods with iron will be dependent not only on the amount of iron added, but also on the overall availability of iron in the diet.

Natural dietary inhibitors that decrease iron absorption include phytates, phosphates,[357] oxalates, and carbonates, which form insoluble precipitates.[789] Substances in tea (tannins),[243] corn,[76, 475, 533] milk, and eggs[76, 145, 560] also reduce absorption. Artificial substances that may interfere with iron absorption include ethylenediaminetetraacetic acid (EDTA), which is added to many foods as a preservative.[200, 291]

The antacids given with various types of therapeutic iron salts reduce absorption of the mineral.[259, 336] No information is available about the effects of antacids on food iron absorption *per se*, although decreased absorption of the nonheme component would be likely. Some types of clay and dirt[552] consumed accidentally or intentionally by subjects with pica will reduce iron absorption. Pancreatic secretions (probably because of the bicarbonate component) also reduce iron absorption.[57]

Iron Losses

Iron balance in the human subject can be upset by iron losses of two sorts: common physiologic losses that exceed iron intake and pathologic iron losses. Iron deficiency may be a consequence of either.

The obligatory physiologic losses in the urine, feces, sweat, and desquamation of cells from epithelial surfaces are discussed in Chapter 4, as is normal menstrual blood loss. Unrecognized menorrhagia is one of the most frequent causes of iron deficiency in women during their reproductive years. Using analyses of subtle changes in hemoglobin concentrations, mean corpuscular hemoglobin concentrations, and serum iron levels, Scandinavian investigators concluded that the upper range of normal menstrual losses was about 60–80 ml.[338] Above this level, bleeding was labeled abnormal or menorrhagic. Factors increasing menstrual flow include genetic factors,[677] multiple childbirth,[676] and the use of intrauterine devices[871] for contraception. Oral contraceptives tend to reduce menstrual losses.[179, 598] As mentioned, pregnancy and delivery contribute to physiologic iron losses. Lactation depletes the maternal iron stores by approximately 0.5–1.0 mg/day,[203] which may be partially counterbalanced by the absence of menses during lactation.

When pathologic causes of iron loss are examined, most of them involve blood loss. In sheer numbers, the menorrhagia suffered by 10–25% of women is the most common form of abnormal blood loss throughout the world. Other obvious froms of blood loss are nosebleeds, wound bleeding, hematuria, and bleeding hemorrhoids. Less overt are the many sources of gastrointestinal bleeding (shown in Table 5-2) that cannot be detected solely by clinical means. In men and post-menopausal women, occult blood loss into the gastrointestinal tract is the most common cause of iron deficiency. The frequency of lesions responsible for this bleeding was estimated by Beveridge.[70] He found that hemorrhoids were responsible in 10% of the cases, salicylate ingestion in 8%, peptic ulceration and hiatus hernia in 7% each, diverticulosis in 4%, neoplasms in 2%, and undefined causes in 16%.

Table 5-2.   Sources of blood loss in the digestive tract[a]

Esophagus	Biliary tract
Varices	Trauma
Hiatus hernia	Cholelithiasis
Stomach	Neoplasm
Varices	Aberrant pancreas
Ulcer	Ruptured aneurysm
Carcinoma	Intrahepatic bleeding
Gastritis	Colon
Leiomyoma	Ulcerative colitis
Small bowel	Amebiasis
Ulcer	Carcinoma
Aberrant pancreas	Telangiectasia
Meckel's diverticulum	Diverticulum
Telangiectasia	Rectum
Polyp	Hemorrhoids
Carcinoma	Ulceration
Regional enteritis	Carcinoma
Helminthiasis	
Vascular occlusion	
Intussusception	
Volvulus	
Leiomyoma	

[a] From Fairbanks et al.[276]

Infestation with hookworm is a common cause of blood loss and iron deficiency in tropical countries.[479] It is estimated that a worm load sufficient to produce 1,000 eggs/g of feces will result in the loss of 2.1 ml blood (0.8 mg iron) per day when *Necator americanus* is the parasite and 4.5 ml blood (1.8 mg iron) per day when *Ancylostoma duodenale* is the agent.[670] Further studies set an upper limit on the degree of parasitism—2,000 eggs/g feces in women, and 4,000–5000 eggs/g in men—that would not undermine the hemoglobin concentrations of a population.[480] Parasitism of greater degrees produced an anemia directly correlated with egg counts because blood and iron losses exceeded the compensatory iron absorption mechanisms.

Other less common forms of abnormal iron loss include blood donation and disorders that produce intravascular hemolysis followed by iron loss as hemoglobin or hemosiderin in the urine. Blood donation results in the loss of about 200 mg iron per unit removed (assuming a 450-ml unit of blood with an average hemoglobin concentration of 14 g/dl). Repeated blood donations at intervals of about two months can be tolerated by many male donors whose body iron stores are depleted,[611] but who avoid anemia by increasing their absorption through a generous intake of dietary iron.[339] Women donors, at risk because of menstrual losses as well as the lesser intake of dietary iron, frequently develop iron deficiency unless they receive iron supplementation.[295]

Iron can be lost through hemosiderinuria resulting from intravascular hemolysis. In intravascular hemolysis, depletion of haptoglobin leads to filtration of hemoglobin dimers through the glomerulus and iron deposition in tubular cells. These cells, laden with hemosiderin, are later excreted into the urine. Although uncommon in the general population, this process may induce profound iron deficiency in patients with paroxysmal nocturnal hemoglobinuria, prosthetic heart valves, atrial myxoma, or malaria. Urinary iron losses from intravascular hemolysis may be as high as 34 mg/day.[274]

An even rarer cause of iron deficiency is pulmonary siderosis, brought on by hemorrhage into pulmonary alveoli. Iron, deposited as hemosiderin, remains in the alveoli and is not resorbed into the body to any major degree; therefore large deposits of iron may be found in the lung although the individual is iron-deficient.

Malabsorption

Compared to blood loss, impaired gastrointestinal absorption of iron is a rare cause of iron deficiency anemia. Two major types of impaired iron absorption are recognized. One results from gastrointestinal surgery and the other from various disorders commonly lumped together under the term "malabsorption syndrome."

Iron deficiency anemia occurs in most patients after total gastrectomy[2] and in 10–50% of patients who have a partial gastrectomy, especially those who undergo the Billroth II surgery that bypasses the duodenum.[36, 340, 383] Typically, the anemia develops progressively during a 10-year period after the operation and, predictably, it is more frequent and severe in menstruating women.[36] Studies with iron radioisotopes indicate that food iron absorption is impaired;[340, 383, 631] the subjects' absorption of medicinal iron was normal but did not increase appropriately in response to the anemia.[38, 340, 724]

In malabsorptive syndromes such as celiac disease caused by gluten enteropathy, epithelial cells are damaged and villi in the proximal small intestine atrophy. Consequently, iron and various other nutrients are not well absorbed.[31, 32] Clinically detectable improvement following treatment with a gluten-free diet results in improved iron absorption in most patients.[31, 144]

DELETERIOUS EFFECTS OF IRON DEFICIENCY

Several tissue changes and metabolic abnormalities have been observed in association with iron deficiency in human beings and in laboratory animals. In some cases the causal relationship of iron deficiency is clear, but in many instances the association is merely inferential. In the following discussion, an effort is made to distinguish between the two.

Specific Tissue Changes

Iron deficiency, despite its wide prevalence and its associated morbidity, rarely causes death. Consequently, information from autopsies is scarce about pathologic tissue changes.[751, 856] More information is available for tissues that can be studied readily through biopsy during life, such as peripheral blood lymphocytes, bone marrow, the oral cavity, and the gastrointestinal tract.[412] These tissues contain rapidly proliferating cells that are more likely to show a lack of iron earlier than cells with a slow turnover.[216]

Bone marrow   The most obvious change in the bone marrow of an iron-deficient patient is the absence of sideroblasts and of stainable iron granules in reticuloendothelial cells as storage iron deposits become mobilized.[303, 347, 657] The erythroid precursors in a hyperplastic marrow increase,[212, 855] but hypoplasia may also be observed, accompanied by increased numbers of reticulum and plasma cells.[68, 131] Poor correlation often exists between the degree of severity of the anemia and the degree of erythroid hyperplasia.[68, 814] Normoblasts with ragged edges and reduced cytoplasm are characteristic of iron-deficient marrow.[855] However, most of these changes are too general or variable to be pathognomonic.

Gastrointestinal tract   Lesions connected with iron deficiency have been described in all portions of the gastrointestinal tract. The predominant lesion is mucosal cell atrophy, which sometimes includes a component of mononuclear cell infiltration. In cells of the oral mucosa, Jacobs[413, 414] found abnormally thin epithelia, increased mitoses in the prickle cell layer, decreased melanin, greater numbers of inflammatory cells in the subepithelia, and keratinization. Similar changes have been described in biopsies of tongues with atrophied filiform papillae.[37] Both lesions are reversed with iron therapy, although the tongue changes require 7–14 days.[37]

The clinical manifestations of these lesions in the oral cavity include angular cheilitis, a sore red tongue, pallor, and often soreness of the buccal mucosa. These changes are not specific, however, because similar abnormalities are exhibited in such disorders as deficiencies of folic acid and vitamin B_{12}.

Additional lesions in the hypopharynx associated with iron deficiency are mucosal atrophy and web formation in the post-cricoid area of the pharynx and upper esophagus. This syndrome, characterized by difficulty in swallowing, is termed sideropenic dysphagia. When the clinical features of gastric achlorhydria and gastric atrophy are involved, especially in middle-aged women, this constellation is termed the Paterson-

Kelly or Plummer-Vinson syndrome.[439, 556, 618, 751, 804] In some patients (5–15%), the disorder has progressed to carcinoma.[5, 170, 397, 410, 415, 868] However, many investigators believe that the causal connection of iron deficiency with the various manifestations of the Plummer-Vinson syndrome is far from explicit.[171, 172, 263, 410, 415 416, 868] Especially troublesome to clearcut explanation is the geographic distribution of the syndrome, which predominates in Northern Europe and the United States. Almost no reports of it exist from Africa, South America, or the Far East, areas where iron deficiency is far more common. Moreover, similar lesions have been described in the absence of any current or past evidence of iron deficiency.[263] Whether or not factors other than iron deficiency predispose to the post-cricoid webs and gastric atrophy is a moot question, particularly because the incidence of Plummer-Vinson syndrome has decreased in recent years.

Gastritis, gastric atrophy, and diminished acid and intrinsic factor secretions are additional abnormalities related to iron-deficiency anemia.[33, 224 415, 417, 484, 856] Histamine-fast achlorhydria has been found in about 50% of patients with iron deficiency anemia,[417, 484, 741] and an even higher incidence of gastritis has been reported.[33, 224, 432, 483] Whether the gastritis and gastric atrophy are a cause or a consequence of the iron deficiency is debatable;[33, 224, 483] either explanation is possible for individual cases. Iron therapy has improved the appearance of the gastric lesions and returned acid secretion to normal in some patients,[33, 224, 417, 484, 741] especially young people,[131, 485] less chronic cases, and patients whose gastric atrophy is not severe. However, recovery is uncommon in patients older than 30.[417, 484, 741] Antibodies to gastric parietal cells circulate in about one-third of patients with iron-deficiency anemia and histamine-fast achlorhydria.[213, 528] Again, the causal relationship of antibodies to iron deficiency is obscure.

Atrophy of small intestinal mucosal cells in children with iron deficiency anemia has been described.[330, 576] Treatment with iron restored these lesions of the mucosal cells to normal. However, Baker[39] criticized the adequacy of the biopsy specimens in these studies[576] because he could detect no marked morphologic abnormalities in 20 patients with severe iron deficiency anemia. The observations of Kimber and Weintraub[444] may be of greater importance. They found intestinal malabsorption of radioiron by severely iron-deficient children and puppies, a condition reversible after iron repletion. They attributed the phenomenon of malabsorption to decreased iron-containing intestinal enzymes.

Integument   The progressive changes in fingernails from brittleness to thinning, flattening, and eventual concavity (koilonychia or spoon nails) are associated with iron deficiency and respond completely to iron therapy.[387] Decreased cystine content of the nails has been reported in individuals with koilonychia.[424] It is not known if iron deficiency interferes with cystine metabolism or if another defect alters its intake or metabolism. Koilonychia is rare in the United States and is no longer a common disorder in other countries. Hair loss has been associated with iron deficiency in women,[348] but such a symptom is not very specific.

Neurologic changes   Various neurologic symptoms have been attributed to iron deficiency, including numbness, paresthesia, and pain, yet all without objective clinical findings. However, papilledema and increased intracranial pressures not due to other causes and that disappear with iron therapy are well documented.[151, 452]

The high concentrations of nonheme iron found in certain parts of the brain (substantia nigra, globus pallidus, and the red nucleus) are of the same order as those in the liver.[341] Some of this nonheme iron is ferritin; the rest is not clearly characterized by either type or function. In the development of iron deficiency, nonheme iron concentrations in the brain do not drop, as do the nonheme hepatic iron stores.[341] Yet a brief period of severe iron deficiency in the young rat results in a deficit of brain iron that persists in the adult animal, despite a subsequent adequate intake of iron.[221]

Nasal mucosal atrophy (ozena)   Ozena, a disorder that occurs with iron deficiency in eastern European countries,[58, 59] is practically unknown in the United States and the rest of the world. The association of atrophic rhinitis with iron deficiency may be fortuitous, as others have reported failure of iron therapy to correct the condition.[569]

Metabolic Defects

No exact measurement of organ or tissue dysfunction exists that clearly divides the effects of iron deficiency from those of anemia. The question of whether or not patients without anemia but with definite depletion of stores and possible depletion of some key tissue iron component have symptoms related to their iron depletion is unresolvable.[69] Clinicians have observed two discordant patterns. Some patients without overt anemia but with depleted iron stores and borderline lowering of serum iron concentration and transferrin saturation exhibit symptoms of profound fatigue and apathy that are alleviated after iron therapy. Yet other people with moderately severe anemia and clearcut iron deficiency may be vigorous, asymptomatic, and completely unaware of their anemia. Faced with these inconsistencies and uncertain about which key iron-containing biochemical systems to study, investigators have been relatively slow to probe the metabolic defects of iron deficiency with the same enthusiasm accorded to the more easily and successfully measured aspects of the anemia.

Biochemical Compounds and Reactions Altered by Iron Deficiency

Heme proteins   Heme proteins, which function in the process of oxidative metabolism, include hemoglobin, myoglobin, the cytochromes, catalase, and peroxidase. Because iron is central to the function of these proteins, its insufficiency would be expected to affect all cells profoundly. In the absence of adequate iron, hemoglobin and myoglobin synthesis is decreased, and the function of the heme proteins is changed quantitatively rather than qualitatively. Much evidence confirms the decrease in hemoglobin in iron-deficient subjects; little is known about the effect of iron deficiency on human myoglobin. Myoglobin concentrations in various muscles of laboratory animals respond differently to iron deficiency. In rat skeletal muscle, myoglobin concentration markedly decreases,[219] but the concentration of myoglobin in the animal's cardiac muscle is unchanged, and absolute amounts of myoglobin appear to increase during iron deficiency.[219] This phenomenon may be related to the workload of the organ, because myoglobin is similarly spared in the diaphragm.

Cytochromes function in the electron transport system. Cytochromes a (cytochrome oxidase), b, and c are located in the cristae of mitochondria and are responsible for transforming cellular energy into adenosine triphosphate. Cytochrome P450 is located in microsomal membranes of the liver and aids in the oxidative degradation of drugs and various metabolites. Cytochrome b_5 is found in many membranes; it functions in a variety of metabolic activities unique to its cell of residence. Levels of cytochrome activity in iron deficiency vary in different organs. Biochemical and histochemical measurements have shown that cytochrome oxidase is reduced in the buccal and intestinal mucosa of iron-deficient patients.[214, 222, 411] In iron-deficient rats, the enzyme is markedly decreased in intestinal mucosa and skeletal muscle, virtually unaffected in brain and heart muscle, and only slightly below normal in kidney and liver.[216] The microsomal cytochromes P450 and b_5 are hardly affected by severe iron deficiency.[218]

Catalase and peroxidase are involved in the reduction of endogenously produced hydrogen peroxide. These enzymes are widely distributed in many cells, and they are usually contained in small organelles, such as peroxisomes of liver cells or granules of leukocytes. Congenital absence or marked deficiency of catalase is remarkably asymptomatic in many people with acatalasia, although others are afflicted with necrotic mouth lesions.[1] Human red blood cell catalase is lowered in iron deficiency anemia;[42, 524, 681] measurements in other tissues have been made in laboratory animals. Rat liver catalase remains normal in iron deficiency.[66]

Metalloflavoproteins Metalloflavoproteins include succinate dehydrogenase, α-glycerophosphate dehydrogenase, and NADH-dehydrogenase of mitochondria, and monoamine oxidase, xanthine oxidase, and aldehyde dehydrogenase in the cell sap. These enzymes all contain nonheme iron and are present in many tissues of the body. Knowledge of their activities in iron deficiency is virtually nil in human studies and meager in animal experiments.[30, 67, 754]

Enzymes with iron as a cofactor   This class of enzymes includes aconitase (citrate or isocitrate hydrolase). They are deficient in whole blood and leukocytes of iron-deficient patients,[753] and are normal or decreased in various tissues of the iron-deficient rat.[65] Hydroxylation of proline and lysine in collagen synthesis is dependent upon a poorly identified iron-requiring enzyme system.[652] The ferrous iron chelating agent α,α^1-dipyridyl interferes with collagen synthesis when given parenterally to normal rats[844] and iron-deficient rats show a similar deterioration.[174] Impaired collagen formation has been proposed as a possible cause for poor wound healing[753] that accompanies iron deficiency. However, it is more likely that the defect is a nonspecific consequence of the disorder.[525]

Other enzyme systems   Enzymatic activities in which iron is neither a component nor a cofactor may be diminished in iron-deficient rats. Becking[52] showed a 50% reduction of the activity of glucose-6-phosphate reductase, phosphogluconate dehydrogenase, and malate dehydrogenase in iron-deficient rats as compared with normal control animals. Iron deficiency led to reduced disaccharidase activity in the brush border of dog intestinal epithelial cells.[384] DNA synthesis in marrow is inhibited in human iron deficiency,[369] and a similar impairment has been observed in HeLa cells[668] treated with the chelating agent desferrioxamine, perhaps because of interference with ribonucleotide reductase.[385]

Mitochondrial Damage

Injury to mitochondria in several tissues of iron-deficient laboratory animals and humans has been demonstrated by electron microscopy and other analytic techniques.[216-218, 318, 427, 665] The mitochondria are uniformly enlarged and radiolucent. In human bone marrow they also appear to be increased. Similar findings in the heart muscle of iron-deficient rats have suggested that greater numbers of enlarged mitochondria may account for the cardiac hypertrophy in excess of that caused by the increased workload imposed by the anemia. Mitochondria from the small intestine of iron-deficient rats are abnormally fragile.[665] Swelling, vacuolation, and breakdown of cristae in mitochondria from lymphocytes of patients with iron-deficiency anemia[427] and in the hepatocytes of severely iron-deficient rats[218] attest to the diffuse abnormalities produced in these important subcellular organelles. However, translation of these morphologic abnormalities into specific functional defects has not been accomplished.

Functional Abnormalities

Although many symptoms of iron deficiency remain unexplained, measurable functional abnormalities associated with iron deficiency have been documented in humans and in experiments with animals. Most of the abnormalities involve complex interactions between a number of organs and biochemical mechanisms. Iron deficiency impinges upon growth, physical and mental performance, reproduction, and susceptibility to infection. Descriptions of these and other conditions are supplied below.

Growth   The effect of iron deficiency on growth rate appears to have been studied much more carefully in laboratory animals than in humans, perhaps because of the difficulty in separating nutritional iron deficiency from other dietary insufficiencies in children. Studies in rats[215, 220, 538] indicate that depressed growth is a late consequence of iron deficiency that appears long after the onset of severe anemia. Two studies found no association between hemoglobin concentrations and weight in children,[49, 141] although addition of iron to the diet of male infants produced an increase in weight gain not seen in female infants. Judisch *et al.*[433] observed two patterns of infant growth. The first was a rapid weight gain that slowed when iron deficiency developed; the second was a normal weight gain that accelerated when iron was supplied. They concluded that iron administration was associated with increased body weight in infancy, but they were uncertain if such a weight gain were desirable.

Physical performance   Although lassitude and tendency to be fatigued easily are common in patients with iron deficiency even without anemia,[69] documentation of decreased physical performance from iron deficiency alone is rare. It has been difficult to define for humans effective endpoints of tests of endurance or maximal response to workloads.[553]

In one study,[225] reduced heart rate following exercise was observed after correction of iron deficiency anemia with iron therapy. In another,[202] pulmonary efficiency was depressed during fairly vigorous exercise in patients with iron deficiency anemia. Respiratory function was improved by iron therapy but not by placebo. Viteri and Torún[805] and Gardner *et al.*[308] showed that the capacity for physical work was impaired even with small (1–2 g/dl) decreases of hemoglobin concentration within the normal range. However, Hallberg and his colleagues produced iron deficiency by phlebotomy in otherwise normal student volunteers and were unable to detect any appreciable differences in the performance of various standardized physical tasks between the phlebotomized subjects and controls (personal communication, L. Hallberg).

In iron-deficient rats running a treadmill, forced and voluntary running time declined to 25–30% below that of normal controls. Performance improved 3 days after iron therapy and was restored to normal 7 days after treatment.[255] In another study,[285] the treadmill running of iron-deficient rats was markedly impaired as compared with normals. The effects of anemia were eliminated by equalizing the hemoglobin concentrations in both groups of animals. Attempts to identify a biochemical defect in skeletal muscle disclosed that the rate of mitochondrial oxidative phosphorylation with α-glycerophosphate as substrate was uniquely responsive to iron replacement and paralleled recovery of treadmill performance in iron-deficient rats treated with iron. Reduced concentrations of cytochromes and myoglobin and low rates of oxidative phosphorylation with other substrates were observed in these iron-deficient rats, but none of the functions improved rapidly after iron administration. Spontaneous basal activity diminished in rats in the presence of iron-deficiency anemia.[313] This reduction was independent of the severity of anemia, and it was rapidly reversed after treatment with iron. In school children studied by Gandra and Bradfield,[304] daily energy expenditure was not dependent upon the degree of anemia. A similar observation was made in studies of iron-deficient and normal children in an orphanage.[39]

Mental performance Efforts to show quantitative reduction in tests of scholastic achievement or other mental performance in relation to iron deficiency have been notoriously difficult to assess.[639] Significantly lower test scores were reported in a group of iron-deficient adolescent boys and girls compared with matched normal control subjects.[828] Another study, reported only in summary form,[391] compared two groups of black children aged 3–5 from low-income areas of Philadelphia. One group was anemic (9–10 g hemoglobin/dl blood), and the other was injected with enough iron to raise the hemoglobin concentration to 12 g/dl. Comparisons of the two groups by a battery of tests indicated that iron-deficient children exhibited strikingly decreased attentiveness, more aimless manipulation, narrower attention span, and less complex and purposeful activity when compared to their normal counterparts. However, no significant differences were revealed by their scores on the Stanford-Binet and Goodenough intelligence tests. Sulzer[748] reported similar conclusions from a different group of psychologic tests. Others[262, 264] have been unable to detect any differences in the performance of psychologic tests in iron-deficient and normal patients even when patients served as their own controls.[262]

An unusual behavioral disturbance of unknown etiology seen in iron deficiency is pica, or the habitual desire for unnatural articles of food. A long list of specific cravings has been recorded, including earth or clay, laundry starch, ice, olives, writing chalk, and cigarette ashes.[276] Pica in adults usually disappears soon after treatment with iron and may recur repeatedly with the return of iron-deficiency anemia. How iron deficiency is related to this craving is completely unknown. A similar behavioral disturbance, breath-holding compulsions, has been attributed to iron deficiency,[389] but no direct cause and effect relationship has been proved.

Observations of decreased monoamine oxidase activity in iron-deficient rats[754, 755] and humans[143, 870] raise the possibility that this enzyme may be responsible for some of the behavioral disorders of iron deficiency. Lack of monoamine oxidase in the brain of an iron-deficient patient might simulate the effect of drugs given to inhibit monoamine oxidase and produce side effects of restlessness and irritability. To date, these relationships are completely conjectural.

Reproduction   Because iron deficiency anemia is especially common in pregnant women, it is surprising that little information if available about the effects of iron deficiency on fetuses or on a mother's ability to carry the pregnancy to term successfully. No available evidence exists that iron deficiency in a population has a major effect on the ability to conceive or on early fetal wastage.[650] The suggestion that iron deficiency in the mother leads to iron deficiency in the infant resulting from depleted iron stores at birth[742] has been refuted.[235, 299, 466, 745, 860]

The fetus apparently has first option on the maternal plasma iron. Placentas from severely iron-deficient mothers weigh more than those from women who received adequate iron during pregnancy.[54] Such hypertrophy might result in greater blood flow to the fetal placenta, but this proposition has not been proved.

Susceptibility to infection   The relationship between human iron status and susceptibility to infection is controversial and has been thoughtfully reviewed by Sussman.[750] Almost all pathogenic microorganisms have an absolute requirement for iron and their interaction with a human host may be affected by the host's iron status. Evidence can be presented that iron deficiency both increases the risk of infection and that it protects against it; the case for neither view is compelling.[750] Considerable additional information is required about the effects of iron on bacterial growth and virulence, host defense, and the interaction between iron and its physiologic binding proteins, transferrin and lactoferrin, before the connection between iron and susceptibility to infection is clarified (see also Chapter 2).

Potential for absorbing other metals   Iron deficiency in the rat enhances absorption of a variety of metals. Manganese absorption is increased in iron deficiency,[242, 543, 637, 765] and neither metal affects the rat's absorption of the other.[765] However, when manganese was perfused into the duodenums of human subjects with varying iron stores, its absorption increased in those with iron deficiency. Conversely, absorption was inhibited in individuals with adequate iron stores.[765] Retention of a tracer dose of manganese-54 10 days after absorption by an iron-deficient patient was no greater than by a patient with normal iron stores.[765]

Absorption of cobalt by rats and humans[637, 767, 769, 798] is increased in the presence of iron deficiency anemia. Both metals are absorbed predominantly in the proximal small intestine of the rat.[766] In both species, increasing the concentrations of either metal inhibits the absorption of the other,[688, 689, 768] suggesting that iron and cobalt share at least part of a common absorptive pathway. These features have been used in a cobalt test for human iron deficiency anemia.[799]

Absorption of cadmium by iron-deficient mice was four- to sixfold greater than by iron-replete mice.[344] Moreover, cadmium toxicosis produces iron deficiency anemia in the mouse and rat,[139, 641] which can be corrected by additional iron. Whether or not human beings with iron deficiency absorb greater amounts of cadmium is not known.

Absorption of lead increases in rats fed iron-deficient diets, as does the toxic quality of the metal.[447, 719] It is unknown if a similar increase in lead absorption occurs in the iron deficiency anemia of children. The frequent association of iron deficiency and lead poisoning in children aged 1–3 years in low-income groups[173, 756] raises the possibility of increased lead absorption as a secondary effect of iron deficiency. However, pica associated with iron deficiency often leads to ingestion of lead from paint and plaster of older buildings in which these children live and clouds the question of whether increased lead absorption or increased availability is at issue.

Absorption of cesium, magnesium, mercury, calcium, and copper was not increased in iron-deficient rats.[637]

6

Acute Toxicity of Ingested Iron

Acute toxicity of ingested iron appears to be related strictly to exposure to therapeutic iron preparations. It is not associated with ingestion of naturally occurring or other commercially produced substances. In the United States, an estimated 2,000 cases of accidental iron poisoning occur annually, with approximately 12 deaths.[210, 838] During an 18-month period, in which 1,645 children who had ingested toxic agents were seen, ferrous sulfate was the agent in 6.2% of the incidents.[326]

In 1947 Forbes pointed out that small children were accidentally poisoning themselves by swallowing oral iron preparations prescribed for adults.[289] Before that report, only sporadic comments were published on acute, potentially lethal poisoning from ingested iron. Numerous case reports and reviews of acute iron toxicity have been published since.[14, 158, 268, 275, 287, 292, 470, 846, 847]

EPIDEMIOLOGY

Acute iron toxicity is found most commonly in children aged less than 5 years, predominately in the 1–2 year age group.[14, 210] Poisoning results when a child encounters the parent's iron medication and, attracted by the superficial resemblance to candy and the taste of the sugar-containing coating, ingests many tablets before discovery. In adults, cases of acute iron toxicity occur almost exclusively because of suicidal intent.[158, 268, 292, 470]

Acute iron toxicity caused by sources other than medicinal iron preparations does not happen because of the large quantities of iron that must be ingested to produce major poisoning. The average human lethal dose is about 200–250 mg iron/kg body weight.[210] Thus the average adult male must swallow 14 g of elemental iron (234 ferrous sulfate tablets). For the average 2-year-old child, 3 g of elemental iron (50 ferrous sulfate tablets) is the average lethal dose.

However, fatalities have been reported from much smaller doses. The ingestion of 1.2-2.4 g ferrous sulfate led to death in a 17-month-old child; but the death was precipitated by peritonitis contracted following surgery to relieve ferrous sulfate-induced pyloric stenosis.[647] Yet a 17-month-old child recovered after swallowing a dose as large as 4.2 g iron.[149] In one review, the range of fatal amounts of ingested iron was 0.96–3.6 g elemental iron, whereas for nonfatal cases it was 0.3–3 g.[14] Although a few adults will develop mild symptoms of gastrointestinal irritation following ingestion of 60–180 mg of medicinal iron, this quantity, the usual therapeutic dose, usually produces no symptoms of overload or acute adverse effects.

CLINICAL STAGES

The features of acute iron poisoning have been well described[14, 838] and may be divided into four phases.

Approximately 30 min after ingestion of the iron, vomiting—which may be bloody—begins in 80% of the patients. More than half the patients become drowsy or lethargic. Diarrhea, often bloody, develops in over 40% of the victims. The pallor, tachycardia, and rapid respirations that often develop may indicate impending shock. A patient may die during this phase, which extends over 6–12 hr.

During the second phase, improvement is noted, stupor clears, and the vomiting and diarrhea subside. This phase lasts 10–12 hr and may be followed by complete recovery or sudden relapse.

Many patients will relapse after the phase of improvement ends. Fever, pneumonitis, shock, coma, convulsions, and death may follow. This phase occurs approximately 20 hr after ingesting the iron.

In those surviving for 3–4 days after iron ingestion, recovery is rapid. In a small number of patients, gastric obstruction related to pyloric stenosis and stricture formation will develop.

Laboratory findings during the first three phases may include marked leukocytosis, evidence of metabolic acidosis, hyperbilirubinemia, and deranged blood coagulation.[275, 846] Serum iron levels vary, yet do have prognostic value. In one report, shock or coma occurred in only 9 of 112 patients (8%) with an initial serum iron value less than 500 μg/100 ml, but in 37% of those with an initial value in excess of 500 μg/100 ml.[838] The onset of coma or shock indicated a serious prognosis.[838, 847] Without treatment, all patients with these signs died. If neither coma nor shock were present, all patients survived. With current modes of treatment, the mortality rate is 11–17%.[838, 852]

PATHOLOGY

In addition to the removal of tissues during surgery,[669] several autopsies have been reported in children and adults who died from acute iron poisoning.[14, 158, 292]

In the stomachs examined, hemorrhagic necrosis and sloughing of areas of mucosa with extension into the submucosa were common. The necrotic surfaces were covered with iron-positive pigment. The mucosa was diffusely congested and infiltrated with polymorphonuclear leukocytes and mononuclear cells. Basement membranes of lymphatics, capillaries, and venules of submucosa and serosa stained iron-positive. Platelet thrombi were numerous in the submucosal capillaries and veins. Changes in the small intestine were the same as described for the stomach but were less severe. However, the most striking changes were in the proximal small intestine. Infarctions, particularly of the ileum,[669] were found to be associated with thrombosis of the mesenteric veins. Autopsies revealed hemorrhagic necrosis of the periportal portion of the liver lobules.[517] Iron stains of these areas showed finely dispersed iron deposits in the portal vein endothelium, Kupffer cells, and periportal reticulum. Other organs exhibited only general conditions of edema, cloudy swelling, and areas of hemorrhage.

PATHOPHYSIOLOGY

Local Events

Hemorrhagic necrosis of the gastrointestinal tract is a direct effect of the interaction of iron and gastric hydrochloric acid. This strongly acid solution containing ferrous and ferric ions, chloride, and usually sulfate, produces coagulation of protein and extensive corrosion. Although the injury to the intestinal tract may be severe enough to lead to scarring, the notion that loss of blood and fluid from the gastrointestinal tract is a major cause of shock and death has been disproved.[846]

Systemic Events

Most of the systemic effects of poisoning are associated with absorption of excessive amounts of iron, resulting in the appearance of plasma iron not bound to transferrin. In experimental situations employing intravenous injections of ionic iron, exceeding the iron-binding capacity has produced such striking symptoms and events in humans as sneezing, nausea, cutaneous flushing, tachycardia, and hypotension.[312, 841] In dogs, excessive iron is rapidly absorbed across intact intestine, where it is responsible for profound metabolic consequences leading to death.[661, 662] Thus, intestinal mucosal damage need not be invoked to explain the excessive absorption of iron or the following systemic events.

Coagulation defects   No systematic research on clotting and iron poisoning has been reported in humans. Studies in animals[846] suggest that alterations in coagulation are secondary to a direct effect of iron on the various proteins involved in blood coagulation. These reactions can be simulated by a direct addition of ferrous sulfate to human plasma.[846]

Metabolic acidosis   Acidosis is a prominent and consistent feature of experimental iron poisoning in dogs and a common development in iron poisoning in children. Three factors are involved with the onset of acidosis. Hydrogen ions are released from the conversion of ferrous iron to the ferric form in the circulatory system, i.e.:

$$1/4\ O_2 + Fe^{2+} + 5/2H_2O \rightarrow Fe(OH)_3 + 2\ H^+$$

Organic acids, such as citric and lactic acids, are accumulated prior to circulatory failure.[661] Electron microscopic studies of ferrous sulfate-induced liver damage show mitochondrial injury that constitutes an anatomic basis for the organic acid accumulation.[857] In addition to the reasons just discussed, lactic and citric acids accumulate as a result of the anaerobic metabolism associated with shock.

Shock   Shock can be an early feature even of severe nonfatal iron poisoning. Therefore, it is more than a terminal event. The peripheral circulatory failure is induced by absorbed iron.[661] Although how the process works has not been characterized fully, it appears that iron directly affects the peripheral vasculature. In dogs given fatal amounts of iron, six alterations were consistently found: reduced cardiac output; increased total peripheral resistance; decreased plasma volume; hemoconcentration; decreased total blood volume; and lowered blood pressure.[661] Using this and other information, Whitten and Brough presented a hypothetical sequence of the pathophysiologic events of acute iron poisoning, but the data on which they based their sequence were insufficient to substantiate the hypothesis.[846]

TREATMENT AND PREVENTION

Details of treatment are beyond the scope of this report. It is based on efforts to rid the victim of unabsorbed iron by using gastric lavage and cathartics. In patients suspected of having ingested near-lethal amounts of iron and in those with serum iron concentrations in excess of the binding capacity, the physician attempts to promote urinary excretion of the extra plasma iron by binding it with a chelating agent, deferoxamine mesylate. Early treatment is most important.[149, 287, 731] The shorter the interval between ingestion of the iron compound and the initiation of therapy, the better the prognosis.

Prevention is the key to this public health problem. Millions of prescriptions are written each year in the United States for iron-containing medications. Moreover, iron compounds are available for purchase without prescription. Because many of the users are women with young children, considerable opportunity exists for iron ingestion and poisoning to occur in that susceptible group. Public education and the necessity for physicians and pharmacists to make their patients aware of the hazard are therefore of the utmost importance.

In summary, no known natural or dietary sources, such as water, food, or any beverage, are likely to result in acute iron toxicity. Acute poisoning from ingested iron comes from accidental ingestion of medicinal iron, mainly by children 1–2 years of age. Iron poisoning may be fatal and early treatment is crucial. Prevention of exposure is the most important aspect of the problem.

7

Chronic Iron Toxicity

In relation to the total body iron content, the amount of exchange that occurs between humans and their environment is small. Disturbances of iron balance are common, but they are almost invariably such that a reduction in the total body iron content results, as discussed in Chapter 5. A major purpose of this chapter is to explore whether or not iron overload does, indeed, result in toxicity.

As the quantities of storage iron (soluble ferritin and insoluble hemosiderin) in the body rise, the ratio of hemosiderin to ferritin increases. Thus an increase in storage iron is called *hemosiderosis* or *siderosis*, and it may be relative or absolute. A relative siderosis is a part of any anemia not caused by blood loss, and merely represents an internal redistribution of the body's iron, with less in the red cell mass and more in stores. Siderosis may also be localized to certain organs; in idiopathic pulmonary siderosis, it is in the lungs, and in paroxysmal nocturnal hemoglobinuria, the kidneys. Such conditions are examples of focal siderosis. When the total iron content of the body is increased, the extra iron is laid down in storage compounds, and it is called absolute siderosis or *iron overload*. The term *hemochromatosis* is applied when the organs containing grossly excessive amounts of storage iron show pathologic evidence of damage, usually fibrosis.

IRON OVERLOAD

Pathogenesis

When the total body iron content is increased, the extra iron enters either through the gastrointestinal tract or by the parenteral route. In some circumstances, both mechanisms may be operative.

Oral iron overload

Excessive Dietary Iron Intake   Although the absorptive mechanism normally adapts itself to the body's need for iron, the actual quantities absorbed nevertheless increase progressively as the intake of absorbable iron rises.[89] Substantial iron overload theoretically can be a function of large amounts of absorbable iron, but the condition is rare except in southern Africa, where alcoholic beverages are prepared in iron drums and pots.[166] The mean iron content of such brews has been found to vary between 40–80 mg/liter and the pH of the beverages is low. In Western countries, an analog sometimes is encountered in wine-drinking alcoholics.[521] However, the iron content of wines is considerably lower than that of the beer consumed by South African blacks; red wines usually contain between 5–6 mg/liter and white wines lesser quantities.[25, 521, 625] Finally, there have been occasional reports of patients who have misguidedly continued to consume iron tonics over extended periods.[164]

There is good reason to question the traditional assumption that dietary iron overload is simply a function of the increased quantities of the metal ingested. Only when the dietary iron is available for absorption is the overload likely to occur. The importance of availability is well illustrated by the situation prevailing among South African blacks. On the basis of stool analyses, the daily intake of many men was found to be between 100–200 mg.[92, 809] Most of this contaminating iron is consumed in maize porridge cooked in iron pots, or in beer made from maize and sorghum fermented in iron drums. Only very small quantities of the iron in the porridge are absorbed. In contrast, the absorption of iron in the beer is good, presumably as a result of the formation of soluble, easily absorbed complexes during the fermentation process.[87] Teff, a grain consumed in Ethiopia, is heavily contaminated with iron but iron overload is not a problem because the extra iron is present in an unabsorbable form or forms.[386]

Excessive Absorption from Diets of Normal Iron Content   Oral iron overload is possible even when the dietary iron content is not inordinate. In individuals with idiopathic hemochromatosis, amounts of iron in excess of body requirements are regularly absorbed, and iron slowly builds up in the body over many years.[89] The metabolic defect responsible for the inappropriate absorption is unknown. Indeed it seems likely that what is called "idiopathic hemochromatosis" may include several separate diseases affecting at different sites the processes by which iron absorption is regulated.[208]

Regardless of the nature of the defect(s) responsible for increased iron absorption, there is good evidence that idiopathic hemochromatosis is inherited. Between 25% and 50% of first-degree relatives exhibit iron overload of varying degree, and siblings are the most commonly affected.[164] The majority show only a modest increase in body iron stores, but some are heavily siderotic. It is rare to find severe iron overload in successive generations, but families exist in which it has been noted.[319] Because different patterns of expression of the disorder have been described, debate on the mode of inheritance has been spirited.[690] However, such questions become irrelevant if it is accepted that the condition is not a homogeneous one, and that any of several defects may be associated with excessive absorption of iron from the gut. The most common form of inheritance appears to be an intermediate one in which the heterozygote manifests minor derangements of iron metabolism, such as a raised plasma iron concentration and a moderate increase in iron stores, and the homozygous state eventually leads to the accumulation of massive amounts of iron.[690, 851] In contrast, in those families in which successive generations have been affected with fully developed hemochromatosis, the mode of inheritance appears to be dominant.[44, 88, 229] In addition, there are reports of patients with the disease who were the offspring of consanguineous marriages, in which neither parent showed any abnormality; presumably the genetic defect in these circumstances was recessive.[228, 229, 601]

Resolution of the uncertainty is further compounded by the fact that the ultimate expression of the gene in an individual can also be influenced by external factors,[164] such as the amount of iron in the diet and physiologic and pathologic blood losses. Severe iron overload thus would be expected to occur less frequently in affected females and to be uncommon in countries like India, in which the quantities of available iron in the diet are small. Conversely, hepatic dysfunction in the iron-loaded liver might be more widespread and more severe in individuals exposed to a hepatotoxin such as alcohol.[286] Whereas a small proportion of patients with alcoholic cirrhosis certainly do show severe siderosis,[321] iron stores are usually normal in alcohol abusers.[514, 516] Iron overload is rare enough in alcoholics outside of southern Africa to suspect that consumption of alcoholic beverages is not the only factor in such subjects. It seems reasonable to suppose that some of them might be heterozygous for a hemochromatosis gene and, had they not been exposed to alcohol, would never have developed a recognizable clinical syndrome.

Parenteral iron overload In several refractory anemias, life is sustainable over long periods through repeated blood transfusions.[81, 84, 89, 150, 448, 521, 607, 697, 718] Because each 500-ml blood transfusion contains more than 200 mg iron, very large quantities can be introduced via this route. In the absence of bleeding, the body's capacity to excrete the extra iron is extremely limited—a few milligrams daily at the most (see also Chapter 4).[323] As a result, massive quantities of storage iron accumulate. Although iron overload is a serious problem in some patients with acquired aplastic anemias, the majority succumb to other complications of bone marrow failure before the stage of massive overload has been reached. Pathologic changes are more common in thalassemia major and other conditions associated with defects in hemoglobin synthesis, including refractory sideroblastic anemia, pyridoxine-responsive anemia, and refractory normoblastic anemia, and they are occasional manifestations of a variety of chronic hemolytic states. There are two reasons for parenteral iron overload. First, the realization that the quality of life in affected individuals is improved by maintaining the hemoglobin at levels closer to normal has encouraged the regular use of transfusion therapy from an early age.[554] Secondly, in anemias like thalassemia major, in which erythroid activity is markedly increased yet ineffective in delivering viable red cells into the circulation, iron absorption from the gut is increased.[45, 99, 269, 710] The iron overload that develops is therefore not simply the result of repeated transfusions, but also includes an oral component. Hence the useful term "iron-loading anemias" coined by Crosby.

Although parenteral overload also could theoretically result from repeated treatment with injectable iron compounds, no published evidence shows that it represents a practical problem.

Prevalence

The prevalence of iron overload in different populations varies not only with the frequency with which certain genes occur in different populations, but with the composition of the average diet. It is not only the dietary iron content that is important, but also its availability for absorption. Meat is a good source of absorbable iron, whereas cereals and other vegetables are poor sources.[533, 781] Therefore, it is not surprising that no reports of clinically manifest oral iron overload have come from India. Differences in prevalence among certain Western countries may represent a subtle interplay between the frequency of genes associated with a tendency to absorb too much iron and the content of easily absorbable iron in the diet. For example, the greater prevalence of clinical hemochromatosis in Australia as compared to the United Kingdom may relate to the greater per capita consumption of meat in Australia. In France, Germany, and Italy, the consumption of red wines could be responsible.[521]

A real difficulty in assessing the prevalence of iron overload in different countries is that quantitative methods for determining the amount of storage iron in tissues have come into general use only recently. The quantitative approach has required either the chemical estimation of the iron concentrations in affected tissues or the measurement of the urinary excretion of iron after administering specific chelating agents. Such a distinction has considerable relevance. Histologically visible iron often is noted in such organs as the liver,[521] and if cirrhosis is present the condition can be labeled hemochromatosis when the quantities of iron are merely moderate. With these reservations, current United States statistics indicate a prevalence of 1 in 20,000 hospital admissions and 1 in 7,000 hospital deaths in one study,[286] and 4 in 100,000 necropsies in another.[276] Much higher, although misleading, figures would be obtained if statistics were collected in diabetic or liver clinics, because the presenting clinical features are so commonly related to diabetes or hepatic dysfunction. Perhaps the most objective figures available are those from a study in which nearly 4,000 liver specimens, the majority obtained from hospital necropsies in 18 different countries, were analyzed chemically for iron.[167] In only three samples were the concentrations more than five times normal and none was anywhere near the figure of 20–50 times normal found in hemochromatosis. Therefore, severe iron overload, at least as judged by necropsy figures, is a relatively uncommon phenomenon in most countries.

As noted, the exceptions are confined to southern Africa, where many populations consume home-brewed drinks high in iron. There have been two attempts to define the prevalence in this region.[134, 166] In one study carried out in Johannesburg, 70% of black adult males and 25% of black adult females dying in hospital were found to have excessive quantities of iron in the liver.[90] Both prevalence and severity increased with age. Although the degree of siderosis was mild or moderate in most subjects, 20% of middle-aged males had concentrations above 2% dry weight, comparable to amounts described in Western countries for subjects with idiopathic hemochromatosis. As exposure to inordinate quantities of iron in South African blacks is largely confined to adults, iron stores are not abnormally high in infants and children; indeed, iron deficiency is not uncommon in these groups.[165]

The prevalence of those anemias most commonly associated with severe iron overload is not known with any certainty. Aplastic anemia is a rare condition, and has a median survival of only about two years; but thalassemia major and several other chronic refractory anemias are compatible with survival for a number of years. These individuals are attracting the attention of investigators interested in the possible long-term effects of severe iron overload.[554] Thalassemia major reaches its greatest prevalence (1 in 400) in Mediterranean peoples, yet most of the current knowledge concerning the effects of the accompanying iron overload has been obtained from studies of small numbers of affected individuals whose parents migrated to the United States, the United Kingdom, or Australia. Such patients have usually been studied more systematically and they tend to have been treated with more intensive regimens of blood transfusions. The prevalence in these countries varies, but it is very low. In the United Kingdom just over 200 patients are thalassemics, most of them immigrants from Cyprus and the northwestern part of India.[554]

Rate of Iron Accumulation

Most patients with idiopathic hemochromatosis develop clinical symptoms in middle age, when body iron content ranges from 20–40 g.[286, 712] This deposit can result from an average positive iron balance of about 2–3 mg daily, but it is doubtful if the accretion is uniformly spread out over the preceding years. The increased stores tend to depress iron absorption, so that by the time clinical manifestations appear the absorption rate is often within the normal range,[850] which is, of course, inappropriate to the iron-overloaded state. After stores have been removed by phlebotomy, the absorption rate rises to between 5–7 mg daily, suggesting that it is initially high and then gradually declines as the stores accumulate.[72, 850] That the clinical manifestations are 10 times more common in men than in women is probably a result of lower iron intakes and greater losses in females.[286]

In dietary siderosis as it occurs in black South Africans, the body iron content builds up from late adolescence until the fifth decade. The quantities of iron in the body of the severely affected individuals are similar to those found in idiopathic hemochromatosis.[403] Limited isotopic studies suggest a positive daily balance of up to 3 mg between ages 20 and 50 years. However, the rate of accumulation may not be uniform.[92]

Transfusional hemochromatosis is of concern in aplastic and hyperplastic anemias.[89] With aplastic anemia, the rate of iron deposition is directly related to the frequency and the number of blood transfusions, although the quantity of iron in the body may be much less than was present in the transfused blood, since thrombocytopenia with resultant blood loss is a frequent complication in such patients. In hyperplastic anemias, such as thalassemia major, good correlation exists between the amount of iron administered in donor blood and the size of the iron stores, but there is also an appreciable contribution from the gut, especially when the subjects are very anemic. As a result, the rate of iron accumulation is greater than in the aplastic group, although the additional load from the excessive absorption varies according to the level at which the hemoglobin is maintained.[554]

Distribution of Iron

In idiopathic hemochromatosis the highest storage iron concentrations are found in the liver and pancreas, where they are between 50 and 100 times normal.[712] In the thyroid the concentration is usually about 25 times normal, and in the heart, pituitary, and adrenals, between 10 and 15 times normal. Lower concentrations of about five times normal are found in the skin, spleen, kidney, and stomach.

The striking histologic feature of the disorder is the widespread presence of the golden-yellow iron storage compound, hemosiderin, in the parenchymal cells of many organs.[247, 712] Deposition of hemosiderin is heavy in hepatocytes and bile duct epithelium, with lesser amounts in Kupffer cells. In the pancreas, hemosiderin deposits are most prominent in the exocrine cells, but are also found in the islets of Langerhans. Epithelial cells of the thyroid, the parathyroid, and the anterior pituitary glands are all affected, whereas in the adrenals the iron deposits usually are confined to the zona glomerulosa. Hemosiderin granules generally are scarce in the testes. Cardiac involvement is a cardinal feature, and deposits are heaviest in the perinuclear region of the muscle fibers of the ventricles.[136, 712] Smaller amounts are present in the atria and conducting system. Considerable attention has been devoted to investigating the deposition of iron in the joints. The synovial lining may be laden heavily with hemosiderin.[252, 345, 712] Electron microscopy has shown that type B cells are involved rather than the macrophage-like type A cells.[696]

Unlike the massive iron deposits in parenchymal cells throughout the body, reticuloendothelial involvement is not prominent; splenic concentrations of iron are only modestly raised,[163, 712] and iron in the reticuloendothelial cells of the bone marrow is often within the normal range.[107]

The distribution of iron in the dietary iron overload of South African blacks has been studied at all stages of its development, because the degree of siderosis in autopsied subjects ranges from mild to massive. The pattern is very different from that observed in idiopathic hemochromatosis. In the earliest stages, an increase in hemosiderin granules is detectable in the hepatic parenchymal and Kupffer cells.[86] With hepatic concentrations of five to ten times normal, the deposits in these cells get denser, and the portal tract macrophages become involved. When iron concentrations reach 20 times normal, heavy deposits are observable in all three sites. Hemosiderin is visible in the spleen from an early age, and with more severe degrees of iron overload, deposits may even exceed those in the liver.[86] Reticuloendothelial involvement is also apparent in the bone marrow, which may contain as much as 10 g storage iron.[303] There is moderately close correlation between the amounts of storage iron in the marrow and the liver.[107] Iron deposits elsewhere in the body are relatively scanty in most subjects. Thus siderosis is largely confined to the hepatocytes and the reticuloendothelial cells.[377, 403]

This pathologic picture is the usual one in black South Africans with iron overload, but in one set of circumstances a different pattern occurs. About 19% of severely siderotic individuals also exhibit micronodular cirrhosis.[97, 403] In such subjects notable deposits are also found in the parenchymal cells of a number of their organs, including the pancreas, thyroid, adrenals, pituitary, and myocardium. The concentrations of iron are comparable to those described in idiopathic hemochromatosis.

The distribution of iron in subjects with parenteral iron overload is influenced by the nature of the underlying anemia and the length of time over which blood transfusions are administered. When the marrow is hypoplastic, the major impact is on the reticuloendothelial cells of the spleen, liver, and bone marrow.[89] This occurs because donor blood is eventually broken down in these sites, and since the plasma iron turnover is reduced, there is little tendency for it to leave. However, if the condition persists over many years, the iron is slowly redistributed and marked parenchymal loading in tissues may result.

In the second group of anemias, the marrow is hyperplastic rather than hypoplastic, which means that external and internal iron kinetics are different from those prevailing in the hypoplastic anemias. This variation affects the distribution of the excess iron. A large proportion of the red cells produced by the overactive erythroid marrow are so defective that they do not even enter the circulation, and those that do have a shortened life span. As a result, erythropoietic activity is great, and the plasma iron turnover markedly increases. Much of the iron released from the catabolized hemoglobin of both these cells and the transfused blood is rapidly returned to the plasma. In addition, iron absorption in such anemias is typically increased despite the developing iron overload. A major contribution from the gut is particularly likely in those conditions in which hemoglobin synthesis is defective, because the red cells superficially resemble those found in iron deficiency anemia. As a result, oral iron therapy may be given mistakenly over extended periods, as illustrated by reports of severe iron overload in subjects with refractory anemias and hypercellular marrows who have received few or no transfusions.[89]

Although the ultimate iron distribution in subjects with refractory anemias and hypercellular bone marrows may be influenced by numerous extrinsic factors, widespread parenchymal deposits of iron generally are inherent in the disorders themselves. Presumably deposition is a function of the rapid iron turnover through the plasma, with subsequent unloading of iron onto the parenchymal cells of various organs. Unloading is almost certainly facilitated by the iron-saturated transferrin bathing the tissues; at least for the liver, iron uptake by hepatocytes is greatly increased under such circumstances. Why the concentrations of iron are greater in some organs than in others is not known, but the rate of blood flow through the organ and the relative number of transferrin receptors could well be responsible. Reticuloendothelial deposits are also prominent in the refractory anemias with hypercellular marrows.[667] Such a feature is not surprising, because iron turnover through these cells is several times normal. That splenectomy in thalassemia major is followed by some redistribution of iron in the liver and an increased amount in hepatocytes is a subject of some dispute.[62, 667, 859]

It is worth considering iron distribution in the liver in more detail, because the one central denominator of all forms of iron overload is involvement of the hepatocytes. Hepatocyte uptake of plasma iron is facilitated by high saturation of transferrin,[190, 426] a factor found in all forms of iron overload except for the dietary form occurring in black South Africans. Another factor in those conditions associated with increased absorption from the gut (i.e., idiopathic hemochromatosis, dietary iron overload, and refractory anemias with increased but ineffective erythroid marrow activity) is that some of the absorbed iron may be deposited directly in hepatocytes from the portal venous blood. Normally such iron is transferrin-bound, but if the percentage saturation of the metal is high and a large amount is absorbed, a good proportion may not be.[841] Any unbound iron is deposited immediately in the hepatic parenchymal cells. In addition, in any anemia characterized by an appreciable degree of intravascular hemolysis, hemoglobin in the plasma, whether bound to haptoglobin or hemopexin, is taken up by hepatocytes;[367] a proportion of such iron might be expected to remain there. Finally, it is possible that some iron deposited in hepatocytes may be derived from the small amount of iron circulating in the plasma as ferritin or bound to a protein other than transferrin.

Associated Pathologic Findings

When reviewing the pathologic findings associated with severe iron overload, it is customary to consider idiopathic hemochromatosis as the prototype. The major feature of the disorder is the presence of fibrosis in heavily siderotic organs, particularly the liver. A fine monolobular cirrhosis is almost invariably found, the lobules typically separated by wide bands of fibrous tissue.[644, 712] However, sometimes a multinodular, postnecrotic type of cirrhosis may develop.[441] In young subjects the fibrosis can be less severe. Carcinoma of the liver is an important late complication, occurring in about 15% of patients.[286] It is usually a hepatoma, but cholangiomas also have been reported.[247] In the pancreas the number of islets is often reduced, and fibrosis and degenerative changes develop in acinar cells.[712] Some fibrosis may also occur in the pituitary, adrenals, and thyroid, but it is not prominent. Testicular atrophy is manifested in the majority of subjects. Dilation and hypertrophy appear in the myocardium; occasionally degeneration of myocardial fibers and fibrosis are reported.[136] The large joints may show calcium pyrophosphate deposition in hyaline and fibrocartilage.[252] Villous hyperplasia and fibrous thickening of the synovium have been noted, as have degenerative changes in the cartilage.[810] These pathologic changes are seen once the clinical features have become manifest; little is known of what occurs during that long latent period, extending over half a lifetime, during which the iron content of the body is slowly increasing. It is clear from studies of affected siblings that hepatic siderosis precedes the onset of fibrosis.[89]

In the dietary iron overload that occurs so commonly in South African blacks a clearer idea of the pathogenesis of iron overload has been obtained, particularly in regard to hepatic changes. The results of several studies indicate that a definite association exists between severe iron overload and significant portal fibrosis or cirrhosis.[86, 90] This association cannot be explained by the prominence of severe siderosis in older subjects, because marked portal fibrosis or micronodular cirrhosis occurs in only 13% of black adults over age 40 in whom the siderosis is minimal, compared to 71% in severely siderotic individuals (iron concentration >2.0% dry weight).[90] Iron loading in the parenchymal cells of other organs is uncommon. However, if deposition occurs, it is usually not accompanied by fibrosis. The only other common pathologic finding in these subjects is osteoporosis, which is especially marked in the spine and may lead to collapse of vertebrae, particularly in the lumbar region.[702] A link between this form of bone disease and severe iron overload has been demonstrated, and chemical analyses of necropsy specimens have shown an inverse correlation between hepatic iron concentration and the mineral density of iliac crest bone.[518]

Associated pathologic changes in subjects with transfusional hemochromatosis are influenced by the nature of the underlying anemia and the age at which it occurs. In an analysis of 20 patients with hypoplastic anemia who had received more than 47 liters of blood (equivalent to 25 g iron), 12 showed increased portal fibrosis, but only one person had fully developed cirrhosis.[89] In contrast, cirrhosis was much more prevalent in 31 patients with refractory anemias associated with hypercellular erythroid marrows. Although these data suggest that iron overload involving the reticuloendothelial system is better tolerated when parenchymal deposits are not present from an early stage, different investigators have used different pathologic criteria, so that any judgments can be only tentative.

More concrete information relating to thalassemia major has become available. Iron concentrations in the organs of five subjects were observed to be in the range described in idiopathic hemochromatosis.[859] All were cirrhotic and 3 showed frank pancreatic fibrosis. Cardiac hypertrophy and hyaline degeneration of myocardial fibers were found, but very little fibrosis. In a study of 41 patients, 39 of whom had thalassemia major, congestive cardiac failure developed in 26 and pericarditis in 19.[266] Necropsies were performed on 11 subjects. Iron deposition was widespread, as was tissue fibrosis, especially in the liver, pancreas, gonads, thyroid, pituitary, and adrenal glands. The heart was dilated and hypertrophied and microscopy revealed large amounts of iron in muscle cells and histiocytes. Myocardial fibers showed focal degeneration and widespread fibrosis. In a study in which 32 liver biopsies were performed on children with thalassemia major, the severity of the fibrosis correlated closely with both the patient's age and hepatic iron concentration.[667] The rate of fibrotic progression was a function of the liver iron concentration; when the concentration was less than 3% dry weight, progression was relatively slow, whereas it accelerated at higher concentrations. Although the severity of the fibrosis correlated with the degree of parenchymal siderosis, its relationship to the reticuloendothelial involvement was closer. An analysis of 207 patients showed that most of the 37 who had died had succumbed at an early age because of inadequate transfusion therapy and/or hypersplenism.[554] The eight older subjects had received an average of 130.2 liters of blood (range,

68.2–188.9), and death was due to cardiac failure in all of them. Despite its presence in the liver and exocrine pancreas, there was no fibrosis in the heart. A child who had received 65.8 liters of blood for congenital aplastic anemia died of myocardial insufficiency, and her heart was examined by electron microscopy.[685] In addition to intracytoplasmic deposits, some iron was consistently present within the nuclei and mitochondria of the myocardium. The mitochondria were swollen and disrupted, and the myofibrils were reduced.

Clinical Features

The clinical presentation in patients with idiopathic hemochromatosis has been fully described.[89, 276, 286, 712] Therefore, this discussion examines the relationship between the iron deposits and the manifestations of the disease. Attention is paid first to the functional disorders that affect those organs in which the highest concentrations of iron occur. Then the possible connection between excessive quantities of iron in the body and certain secondary and indirect effects on body function is considered.

Direct associations between iron overload and clinical manifestations in idiopathic hemochromatosis The major manifestations are skin pigmentation, hepatomegaly, diabetes, cardiac abnormalities, endocrine changes, and arthropathy. Bronzed skin pigmentation is common in most subjects. Excessive melanin is an invariable finding, but hemosiderin deposits are seen in about 50% of patients as well; in the latter group, the skin is a peculiar slate gray.

Symptoms and signs associated with involvement of the liver are nonspecific, and include cachexia, weight loss, and weakness. The liver is almost always enlarged and firm, and signs of hepatic dysfunction, such as palmar erythema, spider angiomata, loss of body hair, and testicular atrophy, may appear. About 50% of patients have splenomegaly, but ascites is rare. The liver disease usually runs a protracted and benign course,[646] except when alcohol abuse complicates matters. Hepatoma eventually supervenes in about 15% of patients, its development suggested by unexplained weight loss, fever, nodular enlargement of the liver, and ascites.[286]

Symptoms related to the onset of diabetes are observed in more than half the subjects at the time of diagnosis,[251, 740] and administration of insulin is usually necessary. It was once thought that the diabetes was ascribable to islet cell failure resulting from local iron deposits, but recent studies have indicated that its etiology is more complex: at least three factors may be involved. Insulin deficiency is undoubtedly one of them, but insulin resistance brought on by the cirrhosis is another.[647] Carbohydrate intolerance is more severe in hemochromatotics than it is in subjects with alcoholic cirrhosis, and this difference is associated with a notably smaller insulin response.[740] In addition, the prevalence of diabetes mellitus unassociated with marked siderosis is high in relatives of subjects with idiopathic hemochromatosis, suggesting that inheritance of traits for diabetes is independent of the inheritance of hemochromatosis.[40, 251]

Cardiac complications are the presenting feature in about 15% of subjects, but as the disease progresses the percentage rises, and they represent perhaps the commonest cause of death. Supraventricular and ventricular arrhythmias of various types have been described, and both left- and right-sided heart failure may develop. The manifestations are usually those of a congestive cardiomyopathy. In subjects under age 40, cardiac manifestations are more frequently the presenting feature and almost always the cause of death, which usually follows within a year of diagnosis unless specific venesection therapy is instituted.

Endocrine changes are common in idiopathic hemochromatosis. The most striking are loss of libido, impotence, amenorrhea, and other evidence of hypogonadism.[103, 286] For example, 70% of 115 patients studied exhibited loss of body hair and testicular atrophy.[811] Although it has been customary to ascribe these changes to hepatic dysfunction, recent hormonal assays indicate that often both the anterior pituitary and the target organ fail.[716] Luteinizing hormone levels were found to be low in most hemochromatotic subjects with testicular atrophy, but normal in individuals with other types of cirrhosis and a similiar degree of gonadal atrophy.[739] It has been established that the low levels of luteinizing and follicle-stimulating hormones were the result of pituitary rather than hypothalamic disturbance.[71] Testosterone concentrations have also been found to be below normal, and they do not rise after specific trophic stimulation. These results are compatible with a primary disturbance of pituitary gonadotrophic function followed by secondary and irreversible failure of gonadal function. Disturbances of other pituitary trophic hormones, however, are uncommon.

Although arthropathy was recognized only recently as a complication of idiopathic hemochromatosis, its symptoms are often present.[695] In half of those individuals with such symptoms, chondrocalcinosis has been noted.[252] Acute attacks of inflammatory synovitis are characterisitic of the condition, but other clinical syndromes have been recognized.[238]

Iron overload from fermented beverages   In studies of this condition in black South Africans, attention has been directed to features characteristic of idiopathic hemochromatosis. Yet it must be emphasized that the degree of siderosis found in the majority of subjects is not as great as in the idiopathic disease, and that even in heavily siderotic individuals the iron is normally confined to the liver and the reticuloendothelial system. Marked hepatic siderosis occurs in a very high proportion of those subjects who develop micronodular cirrhosis.[97, 403] The prognosis for such individuals is poor. The majority succumb within a year to the effects of liver failure and/or portal hypertension. As noted, the iron distribution in many siderotic individuals with micronodular cirrhosis is different from that found in the absence of cirrhosis. In cirrhotic subjects, iron is heavily deposited in the pancreas and other organs; one set of autopsies revealed that as many as 20% had suffered from diabetes mellitus during life.[403] In another study, 7% of black males attending a diabetic clinic had severe siderosis and micronodular cirrhosis.[701] The clinical presentation in such individuals was characteristic: affected patients were usually male, middle-aged, and underweight; their livers were firm and enlarged; and insulin was needed to control the diabetes. Other features characteristic of idiopathic hemochromatosis have not been reported in this variety of dietary iron overload. Specifically, neither cardiac failure nor an endocrinopathy have been demonstrated to be associated with the siderosis. However, certain other manifestations, including ascorbic acid deficiency and osteoporosis, which appear to be indirect consequences of the increased body iron stores, do occur (see below).

Transfusional siderosis   Clinical manifestations similar to those of idiopathic hemochromatosis have been noted, especially when the affected individuals have survived for extended periods. As mentioned, cirrhosis is rare in patients with aplastic anemia, but it develops almost invariably in those with refractory anemias and hypercellular bone marrows who live long enough. Diabetes has been noted in both groups. In 20 patients with aplastic anemia who had received more than 47 liters of blood, 20% had overt diabetes and in another 25%, glucose tolerance was impaired.[89] Overt diabetes mellitus also is frequent among older patients with thalassemia major; and in one study, insulin response to glucose was delayed and/or diminished in six out of eight thalassemic subjects.[467] Two of these had insulin-dependent diabetes, as did three out of eight patients studied elsewhere.[554]

Other manifestations of disturbed endocrine function in patients with thalassemia major are particularly prominent in those patients who survive into the teens.[46] Growth is almost uniformly retarded,[430] but not from growth-hormone deficiency, as growth-hormone responses to a number of stimuli remain normal. It is probable that chronic anemia is the major factor, because growth rates have improved in subjects who have been transfused more intensively.[435, 585] Although clinical evidence of hypogonadism is usual, its cause is uncertain. Levels of gonadotrophic hormones have been reported as being appropriate to the ages of the patients.[148, 457, 554] In contrast, levels of luteinizing hormone were low in four of five adults tested despite low estrogen and testosterone levels.[468] Thyroid and adrenal function are usually normal. Hypoparathyroidism is thought to be a preterminal complication of the disease.[554]

Possible indirect sequelae of iron overload   Certain disorders of a more general biochemical nature than those just discussed may occur in iron overload. Observations were initially made in black South Africans, but such malfunctions may also occur in patients with other conditions.

The first condition that has been noted is ascorbic acid deficiency, which appears in almost all South African blacks having substantial iron overload. Frank scurvy will develop in some.[702] Ascorbic acid deficiency is most prominent in dietary iron overload, but it has also been noted in some subjects with idiopathic hemochromatosis and transfusional siderosis, particularly in patients with thalassemia major.[55, 605, 820] Moreover, when guinea pigs were made siderotic by repeated injections of iron dextran, they became scorbutic even when their diet contained enough of the vitamin to be adequate for the controls.[821]

South African blacks with massive iron overload are not only ascorbic-acid depleted, but frequently osteoporotic.[518, 702] Some conception of the frequency and strength of the association may be gained from the results of a radiologic survey of 110 asymptomatic middle-aged manual laborers. Seventeen were found to be osteoporotic, as evidenced by vertebral body deformity and other signs of decreased bone density.[518] In another combined clinical and histologic investigation, the association between the bone disease and severe iron overload was found to be highly significant, a conclusion further underlined by the finding in a necropsy study of a negative correlation between hepatic iron concentration and mineral bone density.[518] Statistical analysis suggested that age alone was not responsible for the correlation. In addition, a striking correlation exists between clinical scurvy and osteoporosis in this population.[702] Tissue stores of ascorbic acid have been shown to be extremely low in osteoporotic individuals, even when they do not exhibit frank scurvy.[702]

The incidence of osteoporosis in other forms of iron overload has not been accurately documented, but it has been observed in individuals with idiopathic hemochromatosis.[234, 249] Of special interest is a report that beef cattle in an area of New Zealand where the iron content of the water is very high develop a form of siderosis very similar to that seen in black South Africans. In many severely affected animals, osteoporosis of the vertebrae, sternum, and ribs is prominent.[351]

Relationship to Tissue Damage

The particular toxic role of iron in inducing specific manifestations of organ failure associated with severe iron overload is analyzed in this section.

Iron overload and the liver   Hepatic fibrosis is common to all forms of iron overload, and the evidence suggests that the relationship is a causal one. Almost all subjects with idiopathic hemochromatosis have cirrhosis by the time the condition becomes clinically manifest,[286, 712] and in affected but preclinical siblings iron overload precedes the onset of cirrhosis.[88, 851] In dietary iron overload the facts are even more persuasive, because the incidence of serious portal fibrosis or cirrhosis correlates directly with hepatic iron concentrations.[90] This correlation is illustrated in Figure 7-1. Similarly, in patients with transfusional hemochromatosis, especially those with chronic refractory anemias associated with increased but ineffective erythroid activity, the fibrous response appears to accelerate once iron concentration in the liver reaches 3% dry weight.[667] The relative rarity of cirrhosis in subjects with aplastic anemias who have been transfused repeatedly may be a function of two factors.[89] The iron derived from transfused blood primarily goes to the reticuloendothelial system. Because of the reduced marrow activity, relocation of this iron to other sites is slow. Evidently reticuloendothelial iron, even when it is within the liver, is less likely to produce cirrhosis than iron in hepatocytes. The second potentially important factor is time. Most patients with aplastic anemia succumb within a relatively short time, whereas subjects with other forms of refractory anemia may survive for decades.

The evidence that iron is a fibrogenic agent and can cause hepatic cirrhosis appears overwhelming. Yet several questions still remain unresolved concerning the influence of extraneous factors on the genesis of hepatic fibrosis in siderotic subjects. In transfusional siderosis, they may include serum hepatitis and possibly even chronic anoxia. In the other forms of iron overload the most important factor is alcohol: at least 25% of the cases of idiopathic hemochromatosis have a history of excessive

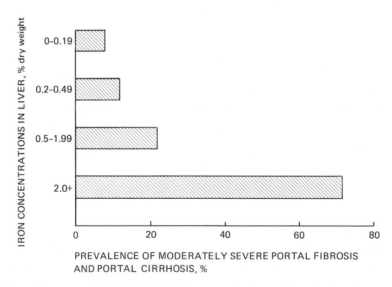

Figure 7-1. Correlation between increasing concentrations of hepatic storage iron and portal fibrosis in black subjects with dietary iron overload. Replotted from data of Bothwell and Isaacson.[90]

alcohol consumption.[286, 521] In the dietary iron overload of South African blacks, alcohol is of even greater importance because the source of the excess iron is a fermented alcoholic beverage.[92, 809] Subjects with the greatest concentrations of liver iron thus tend to be the heaviest drinkers, and the ones whose diets are the least nutritionally adequate. Analysis is further bedeviled by the development of secondary iron overload in a small number of subjects with alcoholic cirrhosis who absorb excessive amounts of iron.[325, 645, 853]

Despite these complications, the ubiquity of hepatic fibrosis in iron overload regardless of pathogenesis supports the notion that iron is central in its production. However, iron probably is only a low-grade fibrogenic agent that requires high local concentrations and protracted exposure to exert its effects. Such an interpretation would be compatible with a necropsy study of dietary iron overload in which 30% of the bodies with hepatic iron concentrations above 2% dry weight showed *no* substantive portal fibrosis.[90] It would also fit with the experimental observation that the siderotic liver is particularly vulnerable to nutritional, metabolic, and toxic insults.[314, 315]

Iron overload and the pancreas   Diabetes is one of the commonest clinical manifestations of idiopathic hemochromatosis: about 80% of idiopathic hemochromatosis patients have diabetes. Its occurrence in cases of dietary iron overload and transfusional hemochromatosis strengthens the concept that iron overload plays a part in its pathogenesis. Evidently, heavy iron deposits in the islet cells of the pancreas lead to reduced insulin output and hence diabetes. However, such a conclusion is not as straightforward as it would first appear. Not only are the iron deposits much less marked in the endocrine than in the exocrine cells,[247, 712] but insulin levels are diminished only in some diabetics with idiopathic hemochromatosis.[251, 740] In addition, familial glucose intolerance and cirrhosis also contribute to the pathogenesis of the diabetes.[40, 251] Insulin levels have not been measured in South African blacks with dietary iron overload, but they have been found to be reduced in patients with thalassemia. In summary, insulin lack, secondary to damage of islet cells by iron, may be one of the factors responsible for diabetes in some patients with iron overload.

Iron overload and the endocrine disorders   Hypogonadism is common in subjects with idiopathic hemochromatosis.[89, 739] Iron deposits are prominent in the anterior pituitary, and hormonal studies have demonstrated low levels of luteinizing and follicle-stimulating hormones[739] unresponsive to hypothalamic-releasing hormones.[71] Testosterone levels are reduced and do not rise after administration of trophic hormones, although testicular iron deposits are scanty.[739, 811] Hypogonadism is found in siderotic thalassemics. Its pattern is similar to that in idiopathic hemochromatosis, although it is unknown whether the primary defect is at the pituitary or the hypothalamic level.[554] The few published results suggest that iron deposits in the pituitary may have a selective effect on function in some patients. The sex trophic hormones are the most susceptible,[468] because the levels of other trophic hormones are usually normal.[148, 457] Interpretation of results is complicated by the sparseness of normal standards for childhood and puberty.[554] The testicular failure should be at least partly ascribable to lack of trophic stimulation.

Iron overload and the heart   The most direct evidence of iron toxicity has been obtained for disorders of cardiac function.[286] Cardiac failure is suggested as the most common cause of death in idiopathic hemochromatosis, as is true for thalassemic subjects whose lives have been prolonged by maintaining an adequate hemoglobin level with repeated blood transfusions.[266, 554] In idiopathic hemochromatosis the cardiopathy is more marked when the clinical disease materializes in youth.[85, 89, 93] Why this should be so is not known, but the phenomenon may relate to the rate of iron accumulation. Fatal cardiopathies in young adults and adolescent subjects with thalassemia who have been heavily transfused are particularly noteworthy, because the age factor tends to discredit the role of variables such as excessive alcohol intake and anemia in the genesis of the cardiopathy. Cardiac complications have not been described in dietary iron overload,[166] and two possible explanations exist for this apparent anomaly. Few subjects accumulate appreciable amounts of iron in the myocardium even when they are heavily overloaded with iron, and those who do succumb rapidly to hepatic failure.

Iron overload and arthropathy   The frequent occurrence of chondrocalcinosis in idiopathic hemochromatosis has led to speculation as to how the iron deposits might contribute to its genesis. It has been suggested that the excessive iron in the joint tissues may inhibit pyrophosphatase.[345] Since this enzyme normally hydrolyzes pyrophosphate to the more soluble orthophosphate, its inhibition promotes pyrophosphate deposition. Complications in joints have not been described in other forms of iron overload.

Iron overload, ascorbic acid deficiency, and osteoporosis   The diet of black South Africans with iron overload certainly does not contain optimal amounts of ascorbic acid, but that alone will not explain the severe tissue depletion of ascorbic acid they exhibit.[702] Ascorbic acid nutrition is adequate in nonsiderotic subjects whose diet is similar except for the beer, and much of the available ascorbic acid is rapidly catabolized in siderotic individuals.[519] When large doses of ascorbic acid are given to such subjects, only small quantities of the vitamin appear in the urine. Instead, more of an end oxidation product of ascorbic acid, i.e., oxalic acid, is excreted. Presumably the rise results from the large deposits of ferric iron, as a similar phenomenon has been observed in other forms of iron overload.[820] Abnormal ascorbic acid metabolism also has been demonstrated in dietary iron overload by tracing the radioactively labeled vitamin.[346] Under such circumstances, the ascorbic acid is also rapidly catabolized, although not to oxalic acid; a large fraction of the carbon-14 label is excreted as carbon dioxide.

The ascorbic acid deficiency accompanying dietary iron overload has several consequences, one of which is interference with the release of iron from reticuloendothelial cells.[494] Such inhibition provides at least a partial explanation for the prominent reticuloendothelial deposits in those subjects. It undoubtedly accounts for their frequently normal plasma iron concentrations, because administering ascorbic acid produces a sharp rise to more appropriate levels.[819] The relatively low degree of transferrin saturation may well be the reason why the unloading of iron onto parenchymal tissues is so rare in this form of iron overload.

That the ascorbic acid deficiency of dietary iron overload is a manifestation of deranged vitamin metabolism and not merely the result of a poor dietary intake has been confirmed by the finding of reduced tissue ascorbic acid concentrations in well-nourished subjects with other forms of iron overload, especially thalassemia major.[820] In addition, administration of large doses of ascorbic acid to patients with an adequate diet is also followed by elevated oxalic acid excretion. Finally, it has been possible to reproduce a similar condition in animals. Guinea pigs made siderotic by injecting iron dextran develop severe ascorbic acid depletion, even when their diets contain adequate amounts of the vitamin.[821]

There is little doubt that iron overload is connected with ascorbic acid deficiency, but the relationship between iron overload and osteoporosis is not so well defined. At least one of the links between the two conditions is ascorbic acid deficiency. Scurvy is known to be associated with osteoporosis in children and laboratory animals, which has been ascribed to the necessity of ascorbic acid for osteogenesis, including collagen, synthesis, osteoblast maturation, and osteoid formation.[802] Bone resorption may be increased in ascorbic acid deficiency; semiquantitative microradiography of the osteoporotic bones of ascorbic acid-depleted guinea pigs has revealed not only the expected diminution in the bone formation surface, but an increase in the bone resorption surface.[821] A similar enlargement of the bone resorption surface has been observed in osteoporotic black South Africans with severe dietary iron overload.[821] In addition, the results of experiments with radioactive calcium suggest that decreased bone formation and increased bone resorption may be present simultaneously.[520] Finally, repletion with ascorbic acid may lead to decreased urinary calcium excretion in such subjects.[520]

Although all this evidence is compatible with the thesis that ascorbic acid deficiency induced by iron overload is responsible for the osteoporosis, especially when coupled with the occurrence of osteoporosis in other forms of iron overload, it is not conclusive. Other factors, such as alcoholism, malnutrition, and associated liver disease, may all play a part. However, their specific roles remain to be elucidated.

Iron overload and porphyria cutanea tarda   Porphyria cutanea tarda is common in siderotic South African blacks.[461] Other sources indicate that mild to moderate iron overload is frequent in porphyria cutanea tarda patients in other countries.[515, 782] The majority of those patients have a history of alcohol abuse, and hepatic fibrosis or cirrhosis are often found.

Possible reasons for the association include the large iron content of some alcoholic drinks,[521] the enhancement of iron absorption that alcohol itself may cause,[168] and the increased iron absorption sometimes observed in alcohol-induced liver disease.[853] Indeed, greater than normal rates of iron absorption have been found in subjects with porphyria cutanea tarda.[782]

However, the association could be causal in the opposite direction, that is, increased iron deposits may affect the pathogenesis of the porphyria. For example, rats given large doses of iron dextran become porphyric more readily when given hexachlorobenzene than do control animals.[758] Of possible relevance, too, is the finding that the rate of *in vitro* uroporphyrin synthesis increases when ferritin and cysteine are added to the system.[460] Although increased amounts of iron may potentiate the metabolic defect in porphyria cutanea tarda, it is not an essential trigger, as the condition has been observed in subjects with normal storage iron concentrations.[515]

Effects of therapy in iron overload   If increased iron deposits are capable of damaging tissues and impairing their function, then removal of the iron would be expected to reverse or at least arrest those changes. In the belief that iron *is* noxious, the treatment of idiopathic hemochromatosis by venesection therapy was started almost 30 years ago.[282] The rationale was a simple one. It was argued that the deficit created ($\pm$ 400 mg iron/liter blood) in the red cell hemoglobin would have to be made good from the iron stores. It was soon confirmed that patients with idiopathic hemochromatosis were able to tolerate the removal of large amounts of blood for protracted periods, so that it was possible to return the body iron content to normal within a year or two.[89] Subsequent studies assessed the effects of such treatment on the function of the most severely affected organs and investigated whether or not survival was prolonged. It has not been possible to institute venesection treatment in siderotic black South Africans, who are typically among the less sophisticated members of the community and unwilling to see large amounts of blood removed. No judgment can thus be made on its therapeutic effects in dietary iron overload. For obvious reasons, venesection has not been practical in subjects with transfusional hemochromatosis. However, attempts have been made to reduce body iron content in such subjects by using specific iron chelating agents such as desferrioxamine and diethylenetriaminepentaacetic acid (DTPA).[24]

It should be stressed that no controlled clinical trial has been performed in which two carefully matched groups of patients with idiopathic hemochromatosis have been monitored, with one being venesected and the other not. Despite this drawback, the uniform consensus nevertheless has been that removing iron is of definite value to the patient. In general, treated patients live longer, and the longer survival is often accompanied by weight gain, reduced skin pigmentation and hepatomegaly, and improved cardiac function.[80, 852]

Gross disturbances in hepatic function are unusual at the time of diagnosis, and therapy usually corrects minor abnormalities. Isolated reports exist of six patients in whom hepatic fibrosis or cirrhosis was judged on the basis of liver biopsy findings to have reverted to normal.[321] In a more systematic study of 85 patients, five who were initially cirrhotic were judged to exhibit only hepatic fibrosis 3–9 years later.[80] However, difficulties in interpreting liver histology on the basis of needle biopsy are well known. Therefore, no final judgment is possible on the degree to which cirrhosis in idiopathic hemochromatosis is reversible by iron removal. There is no evidence that portal hypertension is benefited by venesection therapy.[80] The incidence of hepatoma does not appear to be reduced by treatment. In fact, the disorder may even be greater in treated subjects, although the incidence could be a function of their longer survival rate. Of 45 patients dying from the disease in one study, 13 (29%) succumbed to hepatoma.[80] Ten of the 13 completed a course of venesection therapy, while three died within 6 months of commencing therapy. Three patients died within 3 years and the remaining seven died 4–15 years after the initial venesections.

The degree to which glucose intolerance can be modified by therapy has been carefully documented by Williams and his coworkers, who studied 85 hemochromatotic patients, about two-thirds of whom were diabetic at the time of presentation.[80] The effects of venesection therapy were modest at best. Improved glucose tolerance, designated as a reduction in insulin requirements of at least 12 units/day, or a conversion to oral hypoglycemic therapy, occurred in only 28% of the patients. Fifty-eight percent showed no change in requirements for antidiabetic agents, and 14% needed more insulin after venesection therapy.

Cardiac function often improves after removing excess iron, and venesection treatment undoubtedly has reduced the number of deaths ascribable to the specific cardiopathy.[254, 273, 329, 537, 720] Many patients have not required any maintenance cardiac therapy after the iron has been removed, which further validates the efficacy of such therapy. Timely treatment may also prevent onset of clinical manifestations of cardiac disease in affected individuals.[852]

No relief of hypogonadism or arthropathy has been recorded after venesection treatment.[80]

The most complete statistics on the effects of therapy on survival are those of Bomford and coworkers.[80] They have updated and extended their earlier study, in which they found that the 5-year mortality in 40 phlebotomized patients was only 11%, compared to 67% in a retrospective group of 18 patients.[852] The newer figures are based on 85 phlebotomized patients and a control group of 26 subjects made up of the original 18 patients plus another eight who had refused therapy, died soon after diagnosis, or in whom diagnoses were made at necropsy. The survival at 5 years was 64% in the treated group and 15% in the untreated one. Patients who had died from both groups were matched by age and by the presence of complications, including diabetes, cirrhosis, hepatic failure, and hepatoma. The analysis again yielded a highly significant difference, shown in Figure 7-2. The mean log survival in 45 deceased treated patients was 4.3 years, compared to 2.9 years in the untreated ones. Twenty-two percent of the treated group succumbed to neoplasms other than hepatoma, whereas none did in the untreated group. In another study, retrospective analysis also revealed an improved survival rate in venesected patients, with three deaths in 19 treated indi-

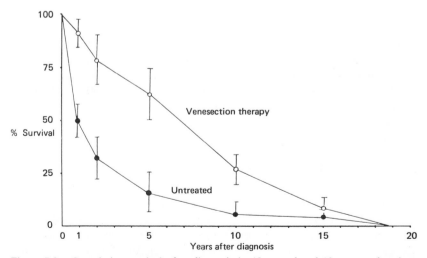

Figure 7-2. Cumulative survival after diagnosis in 45 treated and 18 untreated patients with idiopathic hemochromatosis. The vertical lines at each interval represent ±1 standard error. Reproduced from Bomford et al.[80]

viduals as compared to four deaths in the untreated group of seven subjects.

Data on the effects of chelation therapy in thalassemia major are limited but encouraging. Two groups of children were followed from 5 to 8 years of age. The treated group was given 0.5 g desferrioxamine intramuscularly 6 days a week as well as 2 g DTPA/unit of transfused blood.[46, 667] In both groups, hemoglobin concentrations were maintained between 8 and 15 g/100 ml, and at the start of the trial, the nine children in each group were matched according to age, amount of blood transfused, and hepatic iron concentrations. At the end of 6 years, significant differences were measured in the concentrations of iron in the liver; in the group given chelating agents, mean values were 2.6% dry weight, as compared to 4.2% in the control group. The extent of hepatic fibrosis was also carefully assessed, using an index obtained from camera lucida drawings of liver biopsy specimens. Initially the fibrosis indices overlapped widely, but by the end of the study a highly significant difference had emerged: the fibrosis index had become more marked in the untreated group, yet had remained the same in the chelated group.

These differences might have been more dramatic had more effective chelation therapy been available. Desferrioxamine is effective only after a certain degree of siderosis had been reached, equivalent to 37.6–47 liters of transfused blood.[554] Therefore, it only is possible to contain the iron overload within limits well within the range found in untreated idiopathic hemochromatosis. Despite the limitations, the arrest of the hepatic fibrosis by such therapy[667] provides strong indirect confirmation of the fibrogenic potential of heavy iron deposits in the liver. Other noteworthy clinical observations also came out of this investigation. Puberty was delayed in four out of five control children, but in only one out of four treated patients, and a growth spurt had occurred only in those subjects in whom puberty was not delayed. The incidence of diabetes and its clinical manifestations were similar in each group. In another study in which a larger dose of desferrioxamine was given, some improvement in clinical symptoms and signs was noted—the electrocardiogram and liver function tests showed a return toward normal.[706]

Mild to moderate siderosis commonly accompanies porphyria cutanea tarda. Clinical manifestations subside and porphyrin synthesis decreases when iron stores are depleted by venesection therapy.[655] It is uncertain whether or not only the removal of iron is important for improvement. The subject who embarks on a venesection program may well also stop drinking, which may contribute to the improvement. Iron nevertheless appears to play some key role, because urinary porphyrin excretion has diminished during venesection therapy[267] even in patients who have continued to drink.[512] Biochemical relapse has been induced by replenishing the iron stores.[513]

Animal models of iron overload   The discussion thus far is based on analyses of human disease and its treatment. For a closer examination of the chronic toxicity induced by the metal, an acceptable animal model is necessary. Unfortunately, attempts to create such a model have been unsuccessful.

Much work and many ingenious approaches have not produced hemochromatosis by dietary manipulation. Various species have been fed diets supplemented with iron, and a number of methods have been used to try to increase iron absorption. They have included reducing the phosphate content of the diet, cyclic starvation and feeding, ligation of the pancreatic duct, splenectomy, and administration of such substances as DL-ethionine, polysorbate 20 and D-sorbitol.[89] Despite such maneuvers, body iron concentrations usually have stayed much lower than those found in idiopathic hemochromatosis. In most instances, the iron has been confined to the liver and reticuloendothelial system, although minimal deposits have sometimes been noted in the pancreas, thyroid, and myocardium. Tissue damage, such as hepatic cirrhosis or pancreatic fibrosis, has never been produced by simple dietary iron overload. Success was, however, claimed by workers who gave a choline-deficient, high-fat diet and excessive iron to rats for more than a year.[523] The animals developed fatty livers and progressively increasing cirrhosis; the changes could be prevented by folic acid.[522] However, the heptatic iron concentrations were relatively modest (only six times normal), so that their relevance to hemochromatosis is questionable. Similar doubt must be cast on studies in which cirrhosis has been produced in rats by feeding iron supplements plus carbon tetrachloride,[442] although these experiments do support the contention of Golberg and Smith[315] that iron overload makes the liver more susceptible to other noxious agents. Their view has not gone unchallenged, however, as other workers have been unable to demonstrate such difference when they used ethionine as the second supplement.[250, 858] The discrepancies may be explainable by the finding that iron loading does not potentiate the early "hepatitic" stage of ethionine injury, but rather the second "cirrhotogenic" stage.[612] This second stage involves interaction between excess iron and the cell membrane, but the exact mechanism has not been defined.[442, 857]

Iron has also been administered parenterally to animals as colloidal iron, saccharated iron oxide, iron ascorbate gelatin, iron dextran, and red blood cells. It is usually injected intravenously, but the intraperitoneal and intramuscular routes also have been used. The final body iron content has varied between 0.1 and 3.3 g/kg body weight, and animals were tested any time from 4 weeks to 7 years later.[89] When animals are given very large doses of iron, certain pathologic changes are produced, but they have not been unlike those found in human hemochromatosis. In guinea pigs, for example, subcutaneous injections of inorganic iron to a total dosage of 1 g/kg body weight caused hemorrhagic phenomena in the lungs and adrenals and patchy parenchymal damage of the liver.[599] However, neither cirrhosis nor pancreatic fibrosis was induced. The only morphologic changes in rats subjected to huge doses of parenteral iron were proliferation and hypertrophy of the hepatocytes, ascribed to the induction of protein synthesis.[244] In another experiment, massive iron overload (2.5–3.3 g/kg body weight) was produced in dogs. It caused anorexia, apathy, and weight loss with death after 5–10 months, but autopsies revealed little morphologic evidence of liver injury.[111] Greater success was achieved by Lisboa,[496] who produced cirrhosis in dogs after 4 years by injecting iron intravenously as iron dextran to an accumulated dosage of 3.5–5.8 g/kg body weight. It should be noted that the tissue iron concentrations in these studies were several times greater than those encountered in idiopathic hemochromatosis, and that no pathologic changes appeared even after extended periods during another study in which the iron levels were comparable.[110] Tests for hepatic and cardiac function and for glucose tolerance in these dogs were normal.[110] Serum proteins, liver function as measured by sulfobromophthalin sodium, and glucose tolerance have also been found to be normal in iron-loaded rats and rabbits.[316] Animal experiments in which siderosis has been accompanied by alcohol have also been negative. When rhesus monkeys were given iron dextran intravenously (0.5 g/kg) and were then exposed to alcohol, or alcohol plus a low-protein diet, or carbon tetrachloride, no hepatic fibrosis was apparent in any of the groups even after 110 weeks.[579]

The reasons for the almost uniform failure of parenteral iron loading experiments are several. The first is undoubtedly the predominant reticuloendothelial localization of the injected iron complexes. Although some iron may be redistributed eventually, the amounts taken up by parenchymal cells are usually small. In one study in dogs, for example, almost no iron moved out of the reticuloendothelial cells even after the passage of 7 years.[110] Therefore, the pattern is totally dissimilar to that found in hemochromatosis. If these experiments prove anything, it is the capacity of the reticuloendothelial system to tolerate huge amounts of iron. Another drawback is the necessarily short duration of most such studies, as compared to the length of exposure in the human diseases. Finally, species may well differ in their susceptibility to the noxious effects of excessive tissue iron.

Mechanisms involved in iron toxicity   Despite the largely negative results of attempts to produce experimental hemochromatosis in animals, the cumulative experience in human subjects suffering from iron overload of diverse etiologies strongly suggests that iron *is* noxious to tissues. For it to exert these effects, two criteria must be satisfied: the metal must be present in sufficient concentrations for long periods of time, and it must be present in parenchymal cells. In contrast, the reticuloendothelial cells appear to be admirably equipped to store excessive quantities of the metal. (In the dietary overload encountered in South African blacks, the massive quantities of iron in the spleen and reticuloendothelial cells of the bone marrow do not appear to exert any direct pathologic effects.) Why this striking difference in cellular susceptibility exists is not known. Knowledge of the mechanisms by which iron does damage tissues is itself largely speculative.

Iron has been shown to accumulate in liver cell lysosomes of iron-loaded rats,[780] and liver biopsies from humans with hemochromatosis have also revealed dense deposits of ferritin and hemosiderin in structures tentatively diagnosed as lysosomes.[64, 851] The activity of hepatic lysosomal enzymes has been shown to be markedly increased in patients with various types of iron overload and their lysosomes were strikingly more fragile than normal.[627] Perhaps iron accumulation damages the lysosomal membrane, which then releases acid hydrolases into the cytoplasm and initiates cell damage. These abnormalities return to normal after removing excessive iron from the liver, which suggests that the iron causes them.[627] However, the mechanism by which it injures the lysosomes is not known. The iron deposits may catalyze the formation of free radicals and they, in turn, may damage lysosomes and other subcellular organelles through lipid peroxidation.[627]

The other metabolic abnormality induced by iron overload is depletion of ascorbic acid levels in tissues (see above).[166] Whether chronic subclinical depletion of the vitamin has effects other than scurvy is not known. Lack of its nonspecific antioxidant action may damage susceptible tissues.[554] Serum concentrations of vitamin E (also an antioxidant) have been found to be decreased in thalassemia major.[399]

CARCINOGENESIS, MUTAGENESIS, AND TERATOGENESIS

Carcinogenesis needs to be considered as a complication of iron deficiency, of iron overload states, and of exposure of specific tissues to iron. In connection with epithelial changes in the esophagus of patients with chronic iron deficiency, a few reports have appeared in the older literature of carcinoma of the mid-esophagus.[397] Although this condition has been assumed to be related to iron deficiency, the causal relationship is poorly established. A more definite association has been found between carcinoma of the liver and parenchymal iron overload, described in approximately 15% of patients with idiopathic hemochromatosis.[286] The malignancy may become manifest years after the removal of excess iron;[852] presumably the changes have a long gestation period not reversible by iron removal. A remote possibility exists that the malignancy is caused by some other substance that is also absorbed in excessive amounts by hemochromatotics and which is not removed by phlebotomy. Neoplasms of other organs also are more frequent in idiopathic hemochromatosis.[852] There is no suspicion of local malignancy accountable to gastrointestinal ingestion of inordinate amounts of iron.

Intramuscular injections of iron dextran into animals have produced sarcoma.[314, 511] This finding has implications for the therapeutic use of this substance in treating iron-deficient patients who cannot respond to oral iron. Isolated case reports of local malignant change in such patients give some substance to this.[324] However, iron dextran is now given largely intravenously, which should circumvent the possibility of risk.

Iron has not been reported to be mutagenic or teratogenic.

8

Inhalation of Iron

SOURCES OF EXPOSURE

For Animals

The exposure of animals to inhaled iron is almost exclusively a laboratory phenomenon. Syrian golden hamsters, guinea pigs, mice, and rats are the most commonly used animals in such experiments.[98] Laboratory exposure usually is through natural inhalation or through intratracheal injections of iron oxide particles of 5 μm or less and in single doses as high as 7 mg/m^3 or a 2% solution.[21, 364]

Although some reports[592, 682] suggest that pulmonary fibrosis can occur after inhalation of iron oxide, others studying tumor formation[365, 366, 381] failed to note fibrosis in the lungs when iron oxide served as a vehicle for benzo[a]pyrene and other polycyclic hydrocarbons found in coal tars.[364, 705]

The lungs of animals exposed to ambient iron oxide near steel mills and mines have not been examined for the effects of such exposure *per se*.[714] Cattle exposed to airborne dust near steel plants that is particularly high in fluorides have been autopsied. Their lungs showed anthracosis along with slight to moderate involvement of the regional lymph nodes with carbonaceous material. No evidence of any fibrosis or deposition of iron oxide was found. These reactions decreased as the distance from the steel plants increased.

For Humans

Human exposure to inhaled iron generally comes from an occupational source. The following occupations present risks of inhalation of dust and fumes of iron and its alloys and compounds: iron-ore mining, arc welding, metal (iron) grinding, iron and silver polishing, metal working,[394] sintering, scarfing (personal communication, M. Bundy), and pigment manufacturing and rubber manufacturing.[633]

The Industrial Hygiene Association[181] issued an ambient air quality guide for iron oxide, Fe_2O_3. It stated that no potential acute or chronic hazards were observed in exposed humans and animals even at the highest air concentrations reported. Iron oxide alone[866] is "biologically inert" in that it produces no direct tissue injury in laboratory animals or in humans as a result of exposures in the workplace.

EFFECT AND FATE OF INHALED IRON

In Animals

Pulmonary clearance   When inhaled, iron oxide particles penetrate into the lung parenchyma of mice[381] and may be retained for more than 100 days. The particles are observed inside macrophages collected around terminal bronchioles and lymphoid tissue. Fisher et al.,[288] using Freund's complete adjuvant (FCA) to study clearance of iron-59 oxide in the rat, found that clearance was increased when iron oxide was administered 1–3 days after the use of FCA, because then the response of free macrophages was the greatest. The amount of iron cleared in the first 24 hr as well as the rate of clearance during the later phases were increased as compared with control animals 40 days after exposure. These findings suggest that phagocytic action is important in both early and late clearance phases.

Saffiotti *et al.*[679] reported that a small portion of the iron-oxide dust introduced into the lungs of Syrian golden hamsters remained within the lungs and the tracheobronchial lymph nodes throughout the life span of the animals. Macrophages engulf the iron particles that reach the alveoli, penetrate the alveolar wall with this burden, and ultimately transport the particles through this means. Some of the particles are carried to the trachiobronchial lymph nodes. This process of deposition, retention, and clearance is not unique to iron oxide—it is common for all particles 5 μm or smaller that enter the alveolar spaces. Alveolar macrophages appeared to affect the clearance of radioactive iron dust from cats, rats, and monkeys.[481] The bronchi of all species were almost completely cleared in less than 2 days, whereas the alveoli took 16–28 days in the rats and cats and 280–300 days in the monkeys.[481]

Pulmonary toxicity   Using a scanning electron microscope, Port *et al.*[643] found acute changes in the surface morphology of hamster tracheobronchial epithelium following thrice-weekly intratracheal administration of ferric oxide in 10 doses of 5 mg each. These multiple infusions of ferric oxide particles smaller than 5 μm suspended in 0.9% salt solution produced a loss of ciliated cells and broad areas of abnormal, enlarged, unciliated cells with roughened or wrinkled surfaces. The abnormal cells were thought to be areas of epithelial hyperplasia but their significance remains unknown. This is the first and so far the only report of epithelial changes induced from administration of relatively large doses of ferric oxide. Whether or not lower doses given over a longer exposure time would produce the same response, or whether or not changes would be found in the lower airways of the bronchial tree, is not known. The scanning microscope seems to be a useful tool for exploring reactions to what was previously considered an "inert" dust particle.

Iron as a vehicle   Iron oxide can serve as a vehicle to transport pollutants into the body via the air passages. Amdur and Underhill[21] exposed guinea pigs to open-hearth furnace dust (90% iron oxide) alone and in combination with several concentrations of sulfur dioxide. Open-hearth dust alone produced no detectable respiratory response even when given at a concentration as high as 7 mg/m^3. Neither did it affect the response to any of the sulfur dioxide, which was administered in doses of 1.6–2.6 μg/ml. The findings were similar for an aerosol of iron oxide.[21] Soluble iron salts, however, do potentiate responses to sulfur dioxide.[21]

In Humans

Pulmonary clearance   The fate of iron particles that enter the human lung through occupational exposure has been studied. Lung biopsies of seven welders with siderosis have shown that some iron remains free in the alveoli and bronchioles, but most of it is taken up in macrophages and transported into the lymphatic channels.[565] Much of the incoming dust is collected on the mucus blanket and is expectorated; the dust is recognized easily by the worker because of its rust color.[343]

Injury to the lung   Investigators[489, 501, 502, 659] who studied the mortality patterns of 59,000 steelworkers for 13 years (1953–1966) were unable to find any toxic effect that might be attributable to the inhalation of iron oxide.

In their study of 4,588 crane operators† in steel plants where exposure to ambient iron oxide occurred, Lerer et al.[489] reported that the total mortality experience, as reflected in a relative risk of 1.00, was the same as the mortality experience of workers who had never operated cranes. When analyzed by specific causes, the Caucasian craneman had a 10% deficit mortality risk for lung cancer. That is, 51.3 deaths were predicted and 47 were observed. The relative risk from other respiratory diseases was 0.81 (24 deaths reported versus 28.8 expected), a decrease when compared to other occupations in the same plant. In a long-term study of 9,655 open-hearth workers, mortality from lung cancer was usually less than 1.00 and mortality from nonmalignant respiratory diseases was similar to that of other steelworkers in the same plant, a relative risk of 1.00.[659]

Caucasian crane operators showed an overall excess risk of dying from cardiovascular and renal diseases, with a relative risk of 1.09.[489] The excess risk seemed to be related to particulate exposure, with the greater risk in areas of greater particulate exposure. The mortality sample of non-Caucasian crane operators was too small to determine relative risks

---

†Crane operators were selected because of reports in the literature that they stood an excess risk of dying from lung cancer.

for cardiovascular disease. Open-hearth workers showed a deficit in mortality from cardiovascular diseases with a relative risk of 0.90, which is statistically significant at the 1% level. If particulates, i.e., iron oxide, are related to this phenomena, as proposed in Lerer aand coworkers' discussion of crane operators, then it is exceedingly difficult to explain the deficit in cardiovascular disease mortality in the open-hearth workers studied by Redmond *et al.*[659]

The open-hearth workers had an excess mortality risk (17.6 deaths predicted versus 31 found) from diseases of the digestive system, particularly ulcers. The relative risk was 1.22, which was statistically significant at the 5% level. Heat stress and physical exertion, rather than particulates, may have been at the root of the disorders.

Lowe[506] did not note the influence of iron oxide in inducing chronic bronchitis and emphysema in steelworkers he examined in regard to exposure to atmospheric dusts in the workplace. The disorders were related only to the smoking habits of the steelworkers.

Teculescu and Albu[761] performed several lung-function studies on 14 workers with a mean age of 43 and who had been exposed to pure iron oxide dust for an average of 10 years. They did not find abnormalities compatible with pulmonary fibrosis, although chest radiographs showed nodular opacities. These opacities are symptomatic of pulmonary siderosis.

Siderosis,* the accumulation of iron oxide in the lungs, is a form of pneumoconiosis. Detailed information on the X-ray appearance and pathologic and physiologic responses of the disorder has been published elsewhere.[343, 566] If exposure to pure iron oxide ceases, the dramatic nodulations on the chest X-ray disappear.[565, 684] They would not subside if fibrosis were a feature of the pneumoconiotic nodule of siderosis.

Although the term pulmonary siderosis is reserved for the accumulation of pure iron oxide,[431] it is probable that inhaling pure iron oxide never leads to fibrotic pulmonary changes, whereas inhaling iron plus certain other substances most certainly does. Reports of pulmonary fibrosis in hematite miners[277, 736] indicate that prolonged exposure, usually over 20–30 years, is required, although some signs have been noted after 10-year exposures. It should be noted that the ore to which the miners were exposed contained approximately 10% free silica, a known fibrogenic agent. When pulmonary fibrosis seems to be associated with occupational exposure to iron oxide, careful investigation usually will uncover a simultaneous exposure to free silica. Thus many exposures are to a mixture of materials and these additional particulates (silica) may

---

* In this chapter, siderosis refers solely to pneumoconiosis resulting from the inhalation of iron particles, and not to the cases of excess iron in the blood or tissues discussed in Chapter 7.

indeed produce fibrotic changes in the lungs. The resulting disease is called silicosis.

Some iron compounds are injurious when inhaled.[694] However, the dangerous properties of an iron compound are a function of the ligand with which the iron is coupled. Thus ferric arsenate and ferric arsenite possess the poisonous property of the arsenical ligand. Similarly, iron pentacarbonyl, $Fe(CO)_5$, known as iron carbonyl, is one of the most dangerous of the metal carbonyls, having toxic and flammable properties.

This metal carbonyl is a yellow-brown liquid at ambient conditions and is highly flammable.[104] Its presence may be suspected whenever high partial pressures of carbon monoxide come in contact with iron or steel vessels. Iron carbonyl is formed in gas manufacturing processes, and it (along with nickel carbonyl) must be removed to eliminate soot formation at the ultimate burning point. Iron carbonyl may be formed in illuminating gas that has passed through iron pipes, in water gas, in coal gas stored underground, and in gases containing carbon monoxide held under pressure in steel cylinders. Traces of this carbonyl have been found in gases produced by gasification of oil or refinery gas with steam over nickel catalysts at atmospheric pressure. Liquid iron carbonyl is used as an antiknock agent in some gasolines.

The signs of iron carbonyl poisoning in animals are similar to those of other metal carbonyls and include respiratory distress, cyanosis, tremors, and paralysis of the extremities. Death may be immediate or delayed for a few days. Human exposures similarly produce signs and symptoms similar for the common metal carbonyls. Immediate exposure produces giddiness and headache, and they are occasionally accompanied by dyspnea and vomiting. The signs and symptons may be relieved if the individual is removed to fresh air; however, dyspnea returns 12–36 hr later, with addition of cyanosis, fever, and cough. If the exposure has been to lethal concentrations, death follows in 4–11 days. Pathologic changes include pulmonary hepatization, vascular changes, and degeneration of structure in the central nervous system.

CARCINOGENESIS‡

Boyd et al.[95] studied the mortality of Cumberland iron ore miners through death certificates of 5,811 men from two mining communities who died between 1948 and 1967. Comparison of the iron miners' experience with that of other local men and the relevant national experience led to suspicions of an occupational hazard of lung cancer associated with hematite mining. Earlier, Faulds and Stewart[278] had

---

‡See Cole and Goldman's review of occupational carcinogenesis for a comparison of risks from iron and other metallic dusts and fumes.[178]

reported an increased risk in Cumberland iron ore miners after studying post-mortem findings. During the 20-year period studied by Boyd *et al.*, among all iron miners (underground and surface workers) the 42 lung cancer deaths observed were 50% higher than the number expected (28), an increment statistically significant at the 5% level. However, the number of deaths from other cancers (74) was very close to what was expected (71). Comparing underground and surface workers, the excess of lung cancer was confined to the underground group (observed 36, expected 20.58); there were 6 deaths among the surface workers, and 7.13 had been expected. The higher incidence of lung cancer mortality in the underground miners persisted when tested by the same method against the proportionate mortality determined from the national standards. Post-mortem examinations revealed a high proportion (37%) of oat-cell tumors. This percentage was similar to the figure (43%) found by Saccomanno *et al.*[678] for neoplasms among uranium miners in Colorado. Data on radon concentrations in the Cumberland mines were available for 1968, the year after the completion of the Boyd *et al.* study. In three of the four mines tested, none of the levels—ranging from 30 pCi/liter to 300 pCi/liter—were below 30 pCi/liter, the maximum permissible level recommended by the International Commission on Radiologic Protection. The lung cancer risk in the British miners was much less than that estimated for radium mines in Schneeberg, Germany, Jacymov, Newfoundland, and Colorado. The radon concentration in these latter mines was many times that found in the Cumberland mines. The average was 100 pCi/liter at Cumberland compared to 2,900 pCi/liter in Schneeberg (even after considerable improvements were made); the highest radon concentration measured at the Cumberland mines (320 pCi/liter) was far below the respective measurements of 25,000 pCi/liter and 59,000 pCi/liter for the Newfoundland and Colorado mines, respectively. It may be, therefore, that the limit of a twofold increase in lung cancer estimated for the West Cumberland iron miners is compatible with their lesser exposure to radon in comparison with that of the other groups of miners, for whom risk estimates extend upwards of 10 times normal for the Colorado miners. The exposed rock in the Cumberland has not revealed unusual radioactivity, and it is speculated that the radioactivity is carried by the mine water, which is believed to be its source in the Newfoundland fluorspar mines.[95] Others have thought radon to be the probable causative agent in the development of lung cancer in hematite underground miners.[566]

Ishinishi *et al.*[404] repeatedly instilled iron dusts collected from an open-hearth furnace as well as dusts containing benzo[a]pyrene into the tracheas of albino rats. The suspensions injected were 7.5 mg iron dust, 7.5 mg iron dust plus 1 mg benzo[a]pyrene suspended in 0.2 ml distilled water. The iron dusts and benzo[a]pyrene particles were no larger than 5

μm. The iron dusts contained 52% iron and less than 1% nickel, chromium, and arsenic. No difference in the death rate occurred between the experimental groups (8 out of 90), and the control group (4 out of 30). Both groups had been followed throughout their entire life. After 15 weekly installations spread over about 4 months, the rats exhibited the following tumor incidences. One male rat out of 14 had a malignant tumor and 3 female rats out of 15 had tumors in the iron dust group. In the iron dust plus benzo[a]pyrene group, 8 out of 17 males and 3 out of 9 females had tumors. In the benzo[a]pyrene group, 3 males out of 13 and 3 females out of 14 had tumors. There were no lung tumors in the control group comprised of 12 males and 14 females. Several sarcomas of organs or tissues were detected, but they were also found in the controls. The lung tumor rates were about twice as high in the iron dust plus benzo[a]pyrene group as in either alone. Because the dust from the iron furnace contained small amounts of arsenic, chromium, and nickel (certain compounds of which can act as carcinogens), it may not be possible to conclude with certainty that iron itself has a carcinogenic potency. (This work also reviewed Saffiotti's experiments,[679] which showed that ferric oxide alone did not produce respiratory tumors.)

A link with carcinoma of the lung has not been reported in surface mining operations.[342] Although questions have been raised,[302, 437, 866] it has been generally concluded that neither iron ore dust (hematite) nor ferric hydroxide predispose to malignancy in humans.[331]

Ferric oxide given to hamsters, mice, and guinea pigs by inhalation or by the intratracheal route has not been found to be carcinogenic.[205, 331] Ferric oxide particles can serve as a vehicle for transporting known carcinogenic agents (e.g., benzo[a]pyrene) into the lungs of animals and can produce squamous metaplasia as well as lung cancer under such circumstances.[365, 366, 643, 680, 705, 733] The use of the scanning electron microscope as described by Port et al.[643] may prove to be extremely useful in determining if neoplastic changes in bronchial epithelium occur with exposure to iron.

9

Summary

IRON IN THE ENVIRONMENT

Iron comprises about 5.4% of the continental crust. Its concentration in soils ranges from 0.7% to 4.2%, where it exists primarily as ferric oxyhydroxides and in clay minerals. Crude ores have from 20% to 69% iron, usually in the form of iron oxide minerals. Iron mobility in the regolith generally increases at lower oxidation potentials and pH's in flooded soils and subaqueous sediments where ferrous iron can exist in solution, at lower pH's and in highly moist aerated soils, in the presence of complexing organic compounds dissolved in soil solutions, and with decreasing concentrations of iron-precipitating anions such as sulfide, carbonate, and phosphate. Bacteria also influence the transport of iron by increasing the rate of iron oxidation, producing complexing organic substances or inorganic acids through their metabolism, or oxidizing or utilizing organic parts of chelates to release iron that can then accumulate as ferric oxyhydroxides.

At chemical equilibrium in a well-aerated river, the dominant forms of iron usually are ferric. Unless organic or inorganic complexing agents are present, the solubility of ferric iron cannot be more than a few micrograms per liter at pH's greater than 4, and the solubility at neutral pH is much lower. River water generally contains some particulate ferric oxyhydroxide, however, which may be in the colloidal size range. Iron concentrations in sea water range from a few tenths of a microgram per liter up to about 3 μg/liter. In mildly reducing environments, dissolved ferrous species are dominant and iron solubility at equilibrium may exceed 50 mg/liter at pH 6. Ground waters in many areas of the United States contain appreciable amounts of ferrous iron. Concentrations between 0.5 and 10 mg/liter are common in such areas and larger concentrations occasionally are measured. When these waters are aerated, the iron is oxidized and precipitated as ferric oxyhydroxide. In strongly reduced systems, iron may be retained by ferrous sulfides, which are highly insoluble until the sulfur is oxidized. Metallic iron is chemically unstable in water and many water supplies contain iron that has been derived from corrosion of pipe and other metal surfaces. Organic complexes of ferrous and ferric iron are found in surface and ground waters, and may increase the solubility of iron or influence its reactivity in other ways. Redox and precipitation reactions of iron generally are fast at near neutral pH levels, and chemical equilibrium is attainable. Many of the processes are mediated by microorganisms.

Iron in the atmosphere in remote areas of the United States and United Kingdom is present in year-average concentrations as low as 0.05–0.09 $\mu g/m^3$, whereas in a United States iron and steel center, it has reached a 1-yr maximum average of 12 $\mu g/m^3$ and a 3-month maximum average concentration of 16 $\mu g/m^3$. Atmospheric iron in remote areas is largely soil-derived. The 5-year (1970–1974) median concentration at remote sites in the United States was found to be 0.26 $\mu g/m^3$; at urban sites, 1.3 $\mu g/m^3$, or five times as high. Maximum concentrations tend to occur in the windy season, i.e., in late winter and early spring. Lowest urban iron concentrations are found in coastal states. The iron fraction of soil-derived particulates is roughly 1 part in 100 or 200; the global rate of entry of iron into the atmosphere is of the order of 1,000 metric tons/day.

World production of pig iron and raw steel within the last 5 years has reached 500 million and 700 million metric tons and continues to increase. The United States' contribution, more than half of which comes from Pennsylvania, Ohio, and Indiana, amounts to 17–19% of the world total. Total particulate emission factors for individual manufacturing operations are known to be affected markedly by the type and degree of pollution control employed. Estimation of atmospheric contaminant concentrations resulting from given emission rates is common practice. Estimation of soil quantities raised by the wind is less common, but practicable, and some measure of control through cultivation methods is available. The overall annual average concentration of total airborne particles in the United States decreased from 80 to 66 $\mu g/m^3$ between 1970 and 1974.

The mining and processing of iron or certain materials containing iron minerals affect the environment in several ways. Waste accumulated from iron ore mining presents a problem only when sulfides are present to produce acidic drainage. Extraction of aluminum from bauxite results in wastes containing 1–25% ferric oxide, and these residues amount to about 5.6 million metric tons annually. Most of the wastes are deposited in adjacent lakes. If uncontrolled, emissions from steel mills can contribute appreciable quantities of iron oxides to the air but the only problem demonstrated to date has been soiling. The carrier properties of these particles have not been adequately investigated, and may be of concern. Wastewaters high in iron produced by the steel industry generally are sufficiently treated. A major environmental problem relating to iron-containing minerals involves the decomposition of pyrite and marcasite in coal mines, which can result in acid drainages with high dissolved iron concentrations seeping into streams. Some 17,600 km of streams have been affected by drainage pollution from both active and exhausted underground and surface mines. An additional problem lies in disposing of the voluminous amounts of sludge, primarily ferric oxyhydroxides, that result from the treatment of acidic drainage.

MICROORGANISMS AND IRON

Iron-containing enzymes are ubiquitous in microorganisms (bacteria, yeast, molds, and the microalgae), where they catalyze essential cellular reactions and promote the grand chemical cycles of the biosphere. Aerobic and facultative anaerobic microorganisms are equipped with diverse mechanisms for acquiring the ferric ion. The high-affinity uptake pathway, which involves organic carriers (siderophores) and their corresponding membrane receptors, is presently the best understood. Both siderophore and cognate receptor are under powerful repression by iron. Certain enteric bacteria, viz., *Escherichia coli* and *Salmonella typhimurium*, form receptors for exogenous siderophores, such as the ferrichromes, which they do not themselves synthesize. Citrate may be viewed as a rudimentary siderophore; *E. coli* has an inducible uptake system for ferric citrate, whereas *S. typhimurium* lacks this permease.

The formation of siderophores and matching receptors in pathogenic species may endow the microorganism with a competitive advantage in iron-poor environments, for example, in the tissues of the host. Similarly, in a mixed culture, those species capable of elaborating siderophores may sequester the available iron and thus achieve superiority over organisms lacking this ability—especially if the latter also are unable to transport the siderophore made by their competitor. Metabolically available iron will repress synthesis of surface receptors for siderophores which, in turn, will render the organism resistant to certain bacteriocins and bacteriophages.

Little is known about the possible role of siderophores in conditioning the soil and making iron available to plants. The same is true of the pharmacology of siderophores and their effect upon mammalian iron transport. Introducing increased amounts of dietary iron may change the microflora of the human large intestine.

Rhodotorulic acid and 2,3-dihydroxybenzoic acid are two microbial products that are being evaluted clinically as agents for removing excess body deposits of iron. Desferal, the trade name of desferrioxamine B, a siderophore from *Streptomyces pilosus*, is the drug of choice for treating acute, accidental iron poisoning in young children. Its utility to patients with chronic iron overload is being investigated.

IRON AND PLANTS

Plants require a continuous supply of iron for growth because it is an essential component of many heme and nonheme enzymes and carriers.

Iron concentration is not usually deficient in soils. However, 25–30% of the world's land is calcareous in the surface horizon, and availability of trivalent iron in aqueous solutions of most calcareous soils is inadequate for plant growth. Iron poisoning may also be observed in plants growing on acid soil. For more than 50 years agronomists have tried to change the soil to fit the plant, but no economical way has been found to supply iron to plants on problem soils. It has been appreciated that some plants are able to use iron in calcareous soil (iron-efficient plants) and some cannot (iron-inefficient plants). Iron-efficient plants use nitrate, iron from ferric phosphate, and ferric chelates, and they tolerate other heavy metals better than iron-inefficient plants. This affinity is hereditarily controlled and it is possible to avoid iron chlorosis by genetically tailoring the plant to fit a problem soil.

IRON METABOLISM IN HUMANS AND OTHER MAMMALS

Iron in the human body is largely in the form of hemoglobin within circulating erythrocytes, which serve to transport oxygen. In the plasma, iron is bound to transferrin, which distributes the iron among body tissues according to their needs. Myoglobin, cytochromes, and various other iron-dependent tissue enzymes perform vital functions in all body cells, yet constitute a small fraction of body iron. Surplus iron is deposited as ferritin and hemosiderin, predominantly in reticuloendothelial cells, liver parenchyma, and muscles.

A unique feature of human iron metabolism is the extremely limited external iron exchange. Recent studies indicate a marked limitation in the availability of food iron for absorption. Available iron varies from < 1 to about 5 mg/day, according to the composition of the food ingested. Heme iron is well absorbed, whereas the absorption of nonheme iron is highly variable, depending on the presence and amount of blocking and enhancing substances in the diet. The intestinal mucosa also modifies absorption according to individual needs, and in men this regulation is effective. However, the amount of dietary iron available to women and infants is so low as to threaten iron balance even when absorption is maximal. The iron requirements of pregnancy are so great as to exceed the amount of iron that can be absorbed from an unsupplemented diet.

Despite the many improvements in modern nutrition, iron balance in females and infants remains precarious. Reduction in caloric intake and diminished iron content of food because of decreased contamination during procurement or preparation undoubtedly contribute to the problem. Perhaps the most important cause, however, is that most dietary iron is now derived from vegetable and other nonanimal sources and thus is of very low availability.

Iron nutrition in animals presents few problems because of their higher intake and more efficient absorption. However, the pig is an exception; its rapid growth makes fortification or supplementation necessary.

IRON DEFICIENCY

Iron deficiency in the United States and worldwide is a major health problem. In sheer numbers of people affected, it is one of the most prominent nutritional deficiencies. Iron deficiency is most common in children, in women during their reproductive years, and particularly in the pregnant woman. In adult men, nutritional iron deficiency is rare, and bleeding is the more important cause of the deficiency. The early stages of iron deficiency before the development of overt anemia were once difficult to measure. Accurate tests, including serum ferritin concentration, transferrin saturation, and red cell protoporphyrin content, are now available for evaluating the iron status of populations.

Because iron is essential to oxidative metabolism of all body cells as well as a key element in many enzymatic functions, it is not surprising that iron deficiency produces many structural and functional abnormalities. The abnormalities in red blood cells, gastrointestinal tract, and integument are rather easily demonstrated. Defects within the cell itself (except for altered mitochondrial morphology) are more subtle. The effects of iron deficiency on behavior, mental and physical performance, growth, reproduction, and susceptibility to infection are poorly defined and difficult to separate from the consequences of anemia and other factors. Animal studies indicate that iron deficiency may predispose to increased absorption of other heavy metals and potentiate their toxicity.

ACUTE TOXICITY OF INGESTED IRON

Acute poisoning from ingested iron is unlikely to be encountered from any source other than medicinal iron. No known natural or dietary sources—such as water, food, or any beverage—are likely to cause acute iron toxicity.

CHRONIC IRON TOXICITY

Dietary iron overload rarely, if ever, occurs, but overload may be produced by large volumes of alcoholic beverage high in iron, as happens in black South Africans. Otherwise, iron overload is caused by abnormal mucosal regulation that permits excessive absorption from a diet of normal iron content, or by the parenteral administration of iron, usually in the form of transfused red cells. Conspicuous differences exist in iron distribution in different types of overload; in idiopathic hemochromatosis, deposits are almost exclusively in the parenchymal tissues, particularly in the liver, whereas with transfusions the iron is deposited predominantly in the reticuloendothelial cell. These differences are important, because it is the parenchymal location of iron which is harmful.

The two genetic disorders most commonly associated with iron overload are thalassemia and idiopathic hemochromatosis. Thalassemia major is relatively rare in the United States, affecting only about 5,000 persons. Current United States figures suggest that the prevalence of clinically manifest idiopathic hemochromatosis is of similar magnitude (1 in 20,000 hospital admissions and 1 in 25,000 hospital necropsies). Asymptomatic siblings with increased iron are considerably more frequent; it remains speculative whether such individuals would accumulate sufficient iron to produce full-blown clinical hemochromatosis if exposed to a high dietary intake of iron or excess alcohol.[806]

Chronic parenchymal iron overload produces characteristic clinical manifestations, including hepatomegaly and liver dysfunction, diabetes, cardiac failure, endocrine abnormalities, arthritis, and abnormal skin pigmentation. These symptoms can be arrested or reversed in the individual with idiopathic hemochromatosis through the removal of excess iron by phlebotomy. In the iron overload of thalassemia, chelates have been used to remove excess iron, and evidence has been obtained that this treatment can arrest tissue damage.

Carcinoma of the liver is a complication of idiopathic hemochromatosis and a causal relationship between iron and this malignancy is thought to exist. The intramuscular injection of iron dextran in animals has produced sarcoma, and the association has been reported in a few instances in humans.

INHALATION OF IRON

Ambient air contaminated with iron oxide has not been responsible for a discernible fibrotic response in humans or animals. Laboratory exposure of animals (mice, rats, guinea pigs, and hamsters) to iron oxide produces only an accumulation of iron oxide in the pulmonary parenchyma, without any evidence of fibrosis. Epithelial hyperplasia in the tracheobronchial tree of the hamster has been the only finding noted.

Occupational exposure to iron oxide in the workplace has been found to produce pulmonary siderosis; however, it is not accompanied by fibrosis. No other documented response has been noted. Pulmonary siderosis itself is a benign disorder.

Suspicions of an occupational hazard of lung cancer associated with hematite mining have not been conclusively proved. Whenever an increased risk has been suspected, another more logical cause can be proposed. The probable cause of the increased risk is the presence of radon in the work environment.

10

Recommendations

1. *Additional research on the aqueous chemistry of iron and precipitated iron compounds is needed to improve the treatment of high-iron waters, such as the products of acid mine drainage.*

   The chemical mechanisms and thermodynamics or the precipitation and aging of ferric oxyhydroxides should be studied, as well as the effects of other elements and organic ligands.

2. *Because coal mining can be expected to increase, we should continue to seek the most practical means of preventing or diminishing the formation of acid mine drainage.*

3. *More information is needed on airborne iron concentrations in relation to the iron content of regional soils and dusts deposited in the vicinity of industrial centers.*

   As a contribution to background data on environmental exposure to chemical species, more information is needed on airborne iron concentrations as they relate to the iron content of regional soils, and also to iron deposited on the earth's surface in the vicinity of industrial centers.

4. *Basic research should be encouraged on the adsorption and catalytic properties of particulate iron compounds in the atmosphere to determine their importance in carrying substances or promoting reactions that affect biologic systems.*

   Emphasis should be placed on the surface properties of ferric oxides, as well as on the effect of their mode of formation and history of these properties.

5. *Studies of the relative solubilities of iron compounds in soils should be pursued to aid in determining the influence of different soils on iron availability to plants.*

   In particular, the solubilities of iron incorporated in different clays, metal oxides, and organic chelates, and the effect of incorporating other elements on the solubility of ferric oxyhydroxides in soils should be investigated.

6. *Basic biochemical research on iron should be promoted.*

   In order to achieve the objectives of Recommendations 7–8, research on molecular and cellular aspects of iron metabolism must be continued.

7. *More information is needed on the basic mechanism and regulation of iron transport in microorganisms.*

Induction and derepression mechanisms and microbiologic iron-transport systems need to be examined in relation to microbial virulence and resistance to infection

It must be determined if siderophores supply or deny iron to higher organisms, and the pharmacologic impact of these substances on animals needs to be ascertained.

Siderophore ligands, although virtually ferric-specific, may form complexes with toxic metal ions like plutonium and move such substances through the food chain. The possibility of this problem needs to be recognized and studied.

8. *Clarification of the molecular processes of absorption and transport and their regulation in plants is needed.*

The manner in which an iron-deficient environment induces metabolic changes that enhance iron uptake should be determined. The subjects to be investigated should include the source of hydrogen ions and the source and nature of reductants released by roots.

The movement of ferrous iron into roots and its oxidation should be studied. The internal pathway of iron, including its oxidation to ferric citrate and ultimate utilization in plant tops, also should be scrutinized.

How other metals and chemical compounds affect the iron metabolism of plants should be pursued in greater detail.

9. *The matching of plant species with local soil conditions should be pursued.*

Regional soil and plant tissue testing laboratories could be established for this purpose.

10. *Iron nutrition should be investigated further.*

Methods have been devised to measure iron availability in food, but studies of enhancing and inhibiting factors are required to understand the intricacies of availability. Investigating the availability of iron in water and its effect on the iron balance of populations consuming water high in dissolved iron also would be helpful.

The effect of iron deficiency on absorption of other metals, particularly lead, should be evaluated.

11. *The prevalence of iron deficiency in the United States should be better defined.*

Large-scale population surveys employing a battery of sensitive screening tests for iron deficiency in statistically valid samples of United States subjects are needed.

The relation between iron deficiency and diet should be examined.

12. *The effects of iron deficiency other than anemia should be defined.*
    Particular attention should be given to the importance of the depletion of tissue-dependent tissue enzymes.

    More discriminating tests of functional abnormalities produced by iron deficiency should be designed. Such tests should include observations of behavior as well as examinations of subcellular and organ functions.

    The interaction between iron and its physiologic binding proteins, transferrin and lactoferrin, needs to be elucidated before the connection between iron and susceptibility to infection is clarified.

13. *Improved diagnostic and therapeutic approaches to chronic iron toxicity should be found.*
    The mechanism by which excess iron produces tissue damage should be determined and its carcinogenicity defined. This task would be simplified by the development of a suitable animal model.

    A better means of detecting individuals with excessive iron absorption should be developed.

    More efficient chelating agents and methods of chelate administration should be sought to improve the treatment of patients with iron overload who cannot be phlebotomized.

Appendix
Analysis of Iron in Environmental and Biologic Samples

Many analytic techniques have been developed to measure iron in a variety of materials. More than 1,200 methods exist for the microdetermination of iron in air, water, waste, and biologic samples. Listed in descending order of popularity, the methods were reported as follows: spectrophotometry, 42%, electrochemical, 17%; atomic absorption spectrophotometry, 11%; emission spectrography, 10%; complexometric or volumetric, 10%; radiochemical, 4%; fluorescence, 2%; chromatography, 1.6%; catalytic or kinetic, 1.1%; mass spectrography, 0.8%; microprobe, 0.5%; and chemiluminescence, 0.2%.

ENVIRONMENTAL SAMPLES

Although atomic absorption spectrophotometry (AAS) is a relatively recent development, it has become the method of choice for a large portion of published research during the past 20 years. Many of these less popular methods were created for specific problems and as such are only applicable to such situations. At present, AAS appears to be the preferred technique for determining iron in air, water, and waste samples because of its specificity, speed, and absence of interferences.

Many colorimetric reagents have been proposed for the spectrophotometric determination of iron, but the most common reagents remain phenanthroline and its derivatives and bipyridyl and tripyridyl. Most Environmental Protection Agency (EPA) and Association of Official Agricultural Chemists (AOAC) official procedures employ one of these reagents.

Optical emission and X-ray spectrography have been used to determine iron in environmental samples, but few laboratories have elected to use these techniques routinely. Both techniques are less accurate than the atomic absorption and spectrophotometric procedures. Kopp[454] has reported a coefficient of variation of 7% at a concentration of 64 μg iron/ml for his optical emission technique. An interesting application of X-ray spectrography was reported in which water was filtered through a resin-coated filter paper to retain the metals that were then analyzed by energy-dispersive X-ray fluorescence.[723]

Because of their wide usage at this time, only the atomic absorption and colorimetric procedures will be given a general review. Detailed instructions may be found by consulting the following references:[34, 109,]

[256, 440, 454, 490, 581, 624, 683, 723, 760, 762, 764, 794, 796, 801.] Reliable results can only be obtained from good representative samples properly prepared for analysis. Again, detailed instructions for sample preparation are beyond the scope of this report, but some special procedures will be discussed to provide guidance in selecting a sample preparation technique.

Sample Preparation

Water samples Ferrous and ferric iron are found in water as dissolved solids, colloidal solids, and suspended solids. Under reducing conditions, iron is ferrous, but upon exposure to air or oxidizing materials it will convert to ferric iron. Ferric iron may hydrolize and form insoluble, hydrated ferric oxide. The ferric oxide may adhere to the container wall unless the pH of the sample is low. In general, the word "iron" is used to denote a total of both the ferrous and ferric states.

If ferrous and ferric iron are to be determined separately, a portion of the water sample must be filtered and acidified immediately after collection. Extreme care must be exercised to prevent the oxidation of the ferrous iron and hydrolysis of the ferric ion during filtration.

Colloidal iron tends to adhere to the walls of plastic containers. If suspended iron is to be measured separately, the sample should be shaken vigorously to ensure homogeneity just before removing a portion for analysis. Samples should be analyzed as soon after they are collected as possible to avoid changes in the sample from bacterial growth or oxidation.

If only dissolved iron is to be determined, the water can be filtered through a membrane filter with a pore size of 0.45 μm.[760, 796] The filtrate is used to determine the dissolved iron. However, the pore size of the filter will affect the amount of soluble iron found. Kennedy et al.[440] have demonstrated the variations in the amount of soluble metals in water samples according to the pore size of the filter being used. The suspended material retained on the filter can be measured for suspended iron, or the total iron can be measured on the sample as received and the suspended iron calculated by difference.

Two books may be consulted for specific steps in collecting and preparing water samples.[109, 760]

Particulate air samples It is standard to collect ambient air samples on 20 × 25 cm fiberglass filters and follow the general procedure of the National Air Sampling Network. Environmental samples are collected on a membrane or a fiberglass filter. The collected particulate can be leached from the fiberglass filters by repeated treatments of a hot, 1:1 hydrochloric acid:nitric acid mixture. The leaching solutions are combined and taken to dryness before solubilizing for analysis. The blank filter should always be checked for iron contamination.

Most membrane filters can be destroyed in the hot acid mixture. However, ester-type filters may require repeated ashing with nitric acid to eliminate all the carbon residue of the filter. One type of polyvinyl chloride membrane filter on the market must be given an additional treatment with perchloric acid to complete the destruction of the filter.

Organic membrane filters can be employed for most sampling except when hot metal sparks may strike the filter and burn holes in it. If this is a possibility, a fiberglass filter should be used.

Because of the rigorous acid treatment necessary, it is not possible to differentiate between ferrous and ferric iron in air samples.

Waste material samples†   It is very difficult to collect a representative sample of waste materials because they generally are a mixture of solids and liquids. Rigorous ashing with nitric acid is usually required to destroy all organic materials and solubilize the iron. Sometimes an extra ashing with hydrogen peroxide or perchloric acid is necessary. Perchloric acid is safe to use if a small amount of sulfuric acid is present and if most of the organic material already has been destroyed with nitric acid.

BIOLOGIC SAMPLES

Feeds and Fertilizers

Problems unique to analysis of feeds and fertilizers are usually those of collection and preparation of representative samples. The actual analysis by atomic absorption and/or the various spectrophotometric procedures is straightforward and differs little from procedures used for other materials. Animal by-products used as feeds are analyzable by techniques used for the appropriate animal tissues; accordingly, this discussion of feed analysis will concentrate on plant materials. Much of the material presented here has been excerpted from methods reported in two sources.[27, 328]

Sampling   The AOAC method recommended for sampling bagged fertilizers is to use a slotted tube with a solid cone tip at one end to remove a core of fertilizer diagonally from end to end. For lots of 10 or more bags, at least 10 bags should be sampled. For lots of less than 10 bags, at least 10 samples should be taken, at least one from each bag. For bulk fertilizers, at least 10 cores should be taken from different regions. The samples should be made composite, thoroughly mixed, and reduced to required size by repeated quartering.

---

† Biologic samples are discussed below.

Preparation of samples   The AOAC method for preparing inorganic fertilizers for mineral analysis is as follows. Dissolve 1 g of well-ground sample in 10 ml hydrochloric acid in a 150-ml beaker. Boil and evaporate solution to near dryness on a hotplate; *do not bake*. Redissolve residue in 20 ml of 0.5 *N* hydrochloric acid, boiling gently if necessary. Filter through fast paper into a 100-ml volumetric flask, washing paper thoroughly with distilled water. Dilute to volume with distilled water. If further dilutions are necessary, they can be made with 0.1 *N* hydrochloric acid.

The various organic materials that comprise fertilizers can be prepared for analysis by employing the appropriate dry ashing or wet digestion techniques used for plant samples. With both inorganic and organic fertilizers, precautions should be taken to obtain representative samples and to minimize contamination during all phases of preparation.

Plants

Collection   Plant samples for iron analysis can be collected in paper or polyethylene bags if they have not been treated with talc or other materials used to prevent the bags from sticking together. Aluminum containers are generally not satisfactory for fresh material because they may be corroded by plant exudates. Selecting representative plant samples in the field is critical, especially if mixed forages are being collected. One way of achieving this is to mark off small areas randomly, collect all plant material in each area, and make them into one or more composite samples. Sampling of feed grains is best accomplished in sampling tubes; however, tubes made of noncontaminating materials should be used.

Cleaning and drying   Samples should be cleaned and dried as soon as possible after collection. Contamination of plant samples with soil can be a serious problem and all adherent materials should be removed. Plants can be cleaned by rinsing thoroughly with 0.2 *N* hydrochloric acid or by washing them in detergent solution followed by distilled water rinse. If the same sample is to be used for multiple mineral analyses, care should be taken to avoid leaching losses of the more soluble elements. After thorough cleaning, samples are usually air-dried. The most common procedure is to spread out the plant material in a thin layer and dry it overnight in a forced air oven at 70 C. Drying at higher temperatures may result in decomposition and loss of dry matter.

Grinding and subsampling   Grinding can be a serious source of iron contamination for plant materials. Grinding sometimes can be avoided by analyzing an entire plant; however, if this is not possible, special precautions must be taken to avoid contamination. Using a mortar and pestle or a ball mill made of agate are satisfactory methods. Fragile materials such as dried leaves can often be crushed by hand in a plastic bag. The use of metal-containing mills is often a source of serious but inconsistent iron contamination. Porcelain, flint, or mullite instruments also may contribute to iron contamination.

Ground samples should be dried overnight prior to analysis. After drying, they can be stored in glass or plastic containers with plastic lids. The sample containers should not be more than half-full to allow for thorough mixing before subsamples are taken. Even finely ground materials may segregate because of differences in particle size and density; i.e., stems and leaves often grind differently and thorough mixing is required to assure representative samples.

Dry ashing Most dry ashing procedures in use have been modified to meet the needs of the individual users. The following methods are typical.[328]

Samples (usually 1 or 2 g) are weighed into appropriate containers and placed in a cool muffle furnace. The temperature is raised to 500 ± 50 C for 4–5 hr or overnight. Any unashed organic materials can be removed by wetting the sample with 2 ml of 5 N nitric acid, slowly evaporating to dryness, then returning the sample to a muffle furnace and heating to 400 C for 15 min. The muffle furnace should be cool (< 200 C) when the acid-treated samples are introduced, or violent decomposition may occur. Nitric acid treatment generally is not required. After the sample is thoroughly ashed, it is moistened with a small amount of distilled water and 2 ml of concentrated hydrochloric acid is added. The liquid is evaporated to dryness on a steam plate and allowed to bake 1 hr to dehydrate the silica. The residue is then dissolved in 2.5 ml of 2 N nitric acid and transferred to a volumetric flask. The container should be rinsed several times with hot distilled water; add washings to the volumetric flask. After the sample has cooled and been diluted to volume, allow any silica to settle out or remove it by centrifugation before analysis.

If plant materials that contain large amounts of silica are being ashed, the foregoing procedure sometimes is modified to eliminate potential adsorption of minerals to the silica. Some disagreement exists as to whether or not this problem is serious and the modified procedure may be necessary only when there are large amounts of silica and analysis of the solution for minerals that are present in very small concentrations is anticipated. In this modified method, the samples are treated in the manner previously outlined, including the step of ashing in a muffle furnace. After ashing, the samples are moistened with distilled water, and 5 ml of 48% hydrofluoric acid and 5 drops of concentrated sulfuric acid are added. The samples are then heated gently on a hotplate until the hydrofluoric acid evaporates. The residue is dissolved in 5 ml of 2 N nitric acid and transferred to a volumetric flask with the aid of a stream of hot water.

Wet digestion   In many laboratories, wet digestion of samples with nitric acid and perchloric acids is preferred to dry ashing. Both methods appear to be satisfactory for determining iron. In a typical procedure, a 2-g sample of plant material is weighed into a 250-ml beaker, allowed to stand overnight in 20–31 ml of concentrated nitric acid, heated gently until the initial vigorous reactions have subsided, and slowly evaporated to near dryness. After the sample has cooled, 10 ml concentrated nitric acid, 2 ml concentrated sulfuric acid, and 10 ml of 72% perchloric acid are added. The sample is then returned to the hotplate, covered with a watchglass, and heated until all organic matter is dissolved and the solution is colorless. The coverglass is then removed and the sample is evaporated to near dryness at just below the boiling point. The residue is dissolved in 5 ml of 2 N nitric acid and transferred to a volumetric flask. Silica generally presents few problems when wet digestion is employed.

The use of perchloric acid to digest samples necessitates a number of precautions. The removal of easily oxidized material by pretreatment with nitric acid is essential, as is the addition of sulfuric acid to prevent excessive heating during the evaporation step. Samples should not be allowed to char when evaporating the nitric acid; if charring occurs, more nitric acid should be added and the sample heated until all charred material dissolves. If charred material appears in the sample after perchloric acid has been added, the sample must be cooled and more nitric and perchloric acids added. Heating charred samples in the presence of perchlorates will almost inevitably result in an explosion. Additionally, perchloric acid fumes should only be discharged in hoods designed for perchloric acid use. Alternatively, samples can be digested in Kjeldahl flasks and the acid fumes exhausted via a glass manifold connected to a water aspirator. The problem of perchlorate can be avoided by digestion with sulfuric acid and hydrogen peroxide.[620]

METHOD OF ANALYSIS

Atomic Absorption Spectrophotometry

Atomic absorption spectrophotometry offers the fastest analytic method for determining iron in environmental and biologic samples. (Most plant materials contain enough iron so that no special techniques are required.) It is specific, with generally little or no interference from the other metals in the sample matrix. Four lines in the iron spectrum may be used to analyze solutions, depending upon the iron concentration in the solution to be analyzed. The wavelengths and sensitivities of the four lines are listed in Table A-1.

Although the detection limit (defined as the concentration at which the signal from the analyte is twice the background signal) will vary from

Table A-1. Wavelengths and sensitivities of the iron spectra used in AAS[a]

Wavelength (nm)	Band Pass (nm)	Optimal working range (μg/ml)	Typical sensitivity[b] (μg/ml)
248.3	0.2	2.5–10	0.062
372.0	0.2	25–100	0.55
386.0	0.2	50–200	0.90
392.0	0.2	800–3200	17.0

[a] From Varian Techtron.[801]

[b] Sensitivity is the concentration in an aqueous solution expressed in μg/ml that absorbs 1% of the radiation passing through a cloud of atoms being determined.

instrument to instrument, it is approximately 0.005 μg/ml at 248.3 nm when an air-acetylen flame is used. Citric acid has been reported to suppress the absorbance by up to 50% at a concentration of 200 μg/ml.[801] The nitrous oxide-acetylene flame is supposed to eliminate most interferences.[801]

With the development of the graphite furnace, carbon rod, and tantalum ribbon accessories for atomic absorption spectrophotometers, the detection limit for iron has been reduced by several orders of magnitude. However, it will still vary depending upon the equipment, sample matrix, and specific operating conditions. For instance, Ediger et al., using the HGA-2000 graphite furnace, reported the detection limit for iron in saline water as 0.00002 mg/ml,[256] whereas Bagliano et al., using the HGA-70 graphite furnace, reported a detection limit of 0.001 mg/ml for iron in uranium oxide.[34]

The samples to be analyzed must be free of all suspended particulate to prevent the aspirator system from clogging. The matrix of the calibration standards should be similar to that of the samples to be analyzed. If the sample matrix is very complex, analysis can be accomplished by the method of addition, which will reduce the matrix effects between standards and samples. The iron in the sample can be concentrated and matrix effects reduced by chelating and extracting the iron with ammonium pyrrolidine dithiocarbamate and methyl isobutyl ketone.

Operating procedures are described in the manufacturer's methods manuals and will vary slightly from instrument to instrument. Specific procedures for AAS analysis have been well characterized.[109, 490, 581, 624, 760, 764, 794, 801]

Spectrophotometric Procedures

Review of the literature indicates that many colorimetric procedures have been proposed. However, at this time, either the α,α'- or β, β'-bipyridyls and the phenanthrolines appear to be the reagents of choice.[109, 683, 760, 762] In general, both of these colorimetric reagents have good sensitivity and color stability, obey Beer's law, and can be used over a wide range. The procedures are relatively simple and direct. Both systems react with ferrous iron so that it is possible to determine the ferrous iron directly and the total iron after reduction to the ferrous states. The ferric iron can be calculated by difference.

For accurate results, the iron in the sample must be available readily and not bound up in stable complexes such as those that form with pyrophosphates. Because the reagents react in acidic solutions, the precipitation of hydroxides and phosphates can be prevented.

Conclusions

Colorimetric and atomic absorption spectrophotometry offer satisfactory procedures for measuring iron in a variety of environmental and biologic samples. The specific choice of method must be made by the analyst, based on the type of sample, analysis desired, equipment available for performing the analysis, and any special requirements of the federal agency having jurisdiction. If total iron is to be determined, atomic absorption spectrophotometry is probably the best method. If ferrous and ferric iron need to be measured separately, a colorimetric procedure must be used.

HUMAN AND ANIMAL TISSUES

Assays of interest include measurement of blood iron compounds, primarily used to identify iron deficiency; measurements of iron stores, of utility in evaluating iron deficiency and iron overload; and isotopic measurements, for determining absorption, loss, and internal iron exchange.

The blood of humans and other vertebrates has three important iron compounds. Hemoglobin constitutes most of blood iron and usually is measured colorimetrically by the cyanmethemoglobin method.[209] Plasma transferrin iron is the compound of especial interest because it represents transport iron and reflects the amount of the metal available to body tissues. As a rule, plasma iron is measured colorimetrically.[491] It is not influenced by the presence of hemoglobin, whereas free iron, large amounts of ferritin, and parenteral iron do affect plasma iron. Measuring transferrin-binding capacity is important because the availability of transferrin iron is influenced by the degree of saturation of the iron-binding protein. Colorimetric and radioactive methods have been described.[188] Serum ferritin may be assayed radioimmunometrically;[549] the protein is useful in detecting iron deficiency and overload. Although not an iron compound itself, red cell protoporphyrin increases when the iron supply to the marrow becomes inadequate and therefore serves as an additional gauge of iron deficiency.[462] However, erythrocyte protoporphyrin also rises with lead poisoning and certain unusual intrinsic disorders of porphyrin metabolism.

Reticuloendothelial iron stores can be determined by examining a marrow aspirate for particulate hemosiderin within the reticuloendothelial cells[35] or by determining the amount of iron in marrow particles.[830] Hepatic iron stores may be evaluated by the intramuscular injection of desferrioxamine and subsequent analysis of the urine excreted during the next 24 hr. This technique is particularly effective when parenchymal iron overload is suspected.[349] Serum ferritin is useful in evaluating reticuloendothelial and parenchymal iron stores because its concentration usually parallels the size of tissue iron deposits.[420] Nonheme hepatic iron may be analyzed directly in tissue obtained by biopsy and necropsy.[167]

Much of the information concerning iron turnover has been made possible through the use of iron isotopes. Iron-59, with a half-life of 44 days, is the most convenient and its γ-radiation is detectable by crystal counting *in vitro*, by surface counting, and by total body counting.[89] Iron-55 has a half-life of 4 years; its X-rays are of such low energy that samples must be processed by wet ashing for liquid scintillation counting. Iron-55 and 59 can be counted differentially in the same sample with minimal cross-counting.[253] A crude method for evaluating absorption is the determination of the degree of elevation in plasma iron after orally administering iron to a fasting subject.[258] More accurate determinations can be made by measuring red cell or total-body activity 2 weeks after ingestion of radioiron.[89] Absorption of nonheme and heme iron from complex meals is ascertainable through extrinsic radioiron tags added to the meal.[77, 195]

References

1. Abei, H., and H. Suter. Acatalasemia, pp. 1710–1729. In J. B. Stanbury, J. B. Wyngaarden, and D. S. Fredrickson, Eds. The Metabolic Basis of Inherited Disease. (3rd ed.) New York: McGraw-Hill Book Co., 1972.
2. Adams, J. F. The clinical and metabolic consequences of total gastrectomy. II. Anaemia: Metabolism of iron, vitamin B^{12} and folic acid. Scand. J. Gastroenterol. 3:145–151, 1968.
3. Adamson, J. W., and C. A. Finch. Hemoglobin function, oxygen affinity, and erythropoietin. Ann. Rev. Physiol. 37:351–369, 1975.
4. Agarwala, S. C., C. P. Sharma, P. N. Sharma, and B. D. Nautiyal. Susceptibility of some high yielding varieties of wheat to deficiency of micronutrients in sand culture, pp. 1047–1064. In International Symposium on Soil Fertility Evaluation. Proceedings. Vol. 1. New Delhi: Indian Society of Soil Science, Indian Agricultural Research Institute, 1971.
5. Ahlbom, H. E. Simple achlorhydric anaemia, Plummer-Vinson syndrome, and carcinoma of the mouth, pharynx, and oesophagus in women: Observations at Radiumhemmet, Stockholm. Brit. Med. J. 2:331–333, 1936.
6. Aisen, P. The transferrins (siderophilins), pp. 280–305. In G. L. Eichhorn, Ed. Inorganic Biochemistry. Vol. 1. New York: Elsevier Scientific Publishing Company, 1973.
7. Aisen, P., R. Aasa, and A. G. Redfield. The chromium, manganese, and cobalt complexes of transferrin. J. Biol. Chem. 244:4628–4633, 1969.
8. Aisen, P., and E. B. Brown. Structure and function of transferrin. Prog. Hematol. 9:25–56, 1975.
9. Aisen, P., and A. Leibman. The stability constants of the Fe^{3+} conalbumin complexes. Biochem. Biophys. Res. Comm. 30:407–413, 1968.
10. Ajl, S. J., S. Kadis, and T. C. Montie, Eds. Microbial Toxins. Vol. I. Bacterial Protein Toxins. New York: Academic Press, 1970. 517 pp.
11. Ajl, S. J., S. Kadis, and T. C. Montie, Eds. Microbial Toxins. Vol. II A. Bacterial Protein Toxins. New York: Academic Press, 1970. 412 pp.
12. Ajl, S. J., S. Kadis, and T. C. Montie, Eds. Microbial Toxins. Vol. III. Bacterial Protein Toxins. New York: Academic Press, 1970. 548 pp.
13. Åkeson, Å., G. Biörck, and R. Simon. On the content of myoglobin in human muscles. Acta Med. Scand. 183:307–316, 1968.
14. Aldrich, R. A. Acute iron toxicity, pp. 93–104. In R. O. Wallerstein and S. R. Mettier, Eds. Iron in Clinical Medicine. Berkeley: University of California Press, 1958.
15. Alexanian, R. Urinary excretion of erythropoietin in normal men and women. Blood 28:344–353, 1966.
16. Allgood, J. W., and E. B. Brown. The relationship between duodenal mucosal iron concentration and iron absorption in human subjects. Scand. J. Haematol. 4:217–229, 1967.
17. Ambler, J. E., and J. C. Brown. Cause of differential susceptibility to zinc deficiency in two varieties of navy beans (*Phaseolus vulgaris* L.) Agron. J. 61:41–43, 1969.

18. Ambler, J. E., and J. C. Brown. Iron-stress response in mixed and monocultures of soybean cultivars. Plant Physiol. 50:675–678, 1972.

19. Ambler, J. E., and J. C. Brown. Iron supply in soybean seedlings. Agron. J. 66:476–478, 1974.

20. Ambler, J. E., J. C. Brown, and H. G. Gauch. Sites of iron reduction in soybean plants. Agron. J. 63:95–97, 1971.

21. Amdur, M. O., and D. W. Underhill. Response of guinea pigs to a combination of sulfur dioxide and open hearth dust. J. Air Pollut. Control Assoc. 20:31–34, 1970.

22. American Chemical Society. Cleaning our Environment. The Chemical Basis for Action. Washington, D.C.: American Chemical Society, 1969. 249 pp.

23. American Iron and Steel Institute. Annual Statistical Report. (1975) Washington, D.C.: American Iron and Steel Institute, 1976. 98 pp.

24. Anderson, W. F., and M. C. Hiller, Eds. Proceedings of a Symposium. Development of Iron Chelators for Clinical Use. DHEW Publ. (NIH) 76-994. Bethesda, Md.: U.S. Department of Health, Education, and Welfare, National Institutes of Health, 1976. 277 pp.

25. Anguissola, A. B. The nutritional value of wine as regards its iron content, pp. 71–74. In L. Hallberg, H.-G. Harwerth, and A. Vannotti, Eds. Iron Deficiency: Pathogenesis, Clinical Aspects, Therapy. New York: Academic Press, 1970.

26. Aristovskaya, T. V., and G. A. Zavarzin. Biochemistry of iron in soil, pp. 385–408. In A. D. McLaren and J. Skujiņš, Eds. Soil Biochemistry. Vol. 2. New York: Marcel Dekker, 1971.

27. Association of Official Analytical Chemists. Official Methods of Analysis. (12th ed.) Washington, D.C.: Association of Official Analytical Chemists, 1975. 1094 pp.

28. Atkin, C. L., and J. B. Neilands. Rhodotorulic acid, a diketo-piperazine dihydroxamic acid with growth-factor activity. I. Isolation and characterization. Biochemistry 7:3734–3739, 1968.

29. Aung-Than-Batu, U Hla-Pe, Thein-Than, and Khin-Kyi-Nyunt. Iron-deficiency in Burmese population groups. Amer. J. Clin. Nutr. 25:210–217, 1972.

30. Awai, M., and E. B. Brown. Examination of the role of xanthine oxidase in iron absorption by the rat. J. Lab. Clin. Med. 73:366–378, 1969.

31. Badenoch, J., and S. T. Callender. Effect of corticosteroids and gluten-free diet on absorption of iron in idiopathic steatorrhea and coeliac disease. Lancet 1:192–194, 1960.

32. Badenoch, J., and S. T. Callender. Iron metabolism in steatorrhea. The use of radioactive iron in studies of absorption and utilization. Blood 9:123–133, 1954.

33. Badenoch, J., J. R. Evans, and W. C. D. Richards. The stomach in hypochromic anaemia. Brit. J. Haematol. 3:175–185, 1957.

34. Bagliano, G., F. Benischek, and I. Huber. Application of graphite furnace atomic absorption to the determination of impurities in uranium oxides without preliminary chemical separation. Atom. Absorpt. Newslett. 14:45–48, 1975.

35. Bainton, D. F., and C. A. Finch. The diagnosis of iron deficiency anemia. Amer. J. Med. 37:62–70, 1964.

36. Baird, I. M., E. K. Blackburn, and G. M. Wilson. The pathogenesis of anaemia after partial gastrectomy. I. Development of anaemia in relation to time after operation, blood loss, and diet. Q. J. Med. 28:21–34, 1959.

37. Baird, I. M., O. G. Dodge, F. J. Palmer, and R. J. Wawman. The tongue and oesophagus in iron-deficiency anaemia and the effect of iron therapy. J. Clin. Path. 14:603–609, 1961.

38. Baird, I. M., and G. M. Wilson. The pathogenesis of anaemia after partial gastrectomy. II. Iron absorption after partial gastrectomy. Q. J. Med. 28:35–41, 1959.

39. Baker, S. Discussion, pp. 142–144. In H. Kief, Ed. Iron Metabolism and Its Disorders. New York: American Elsevier Publishing Co. Inc., 1975.

40. Balcerzak, S. P., D. H. Mintz, and M. P. Westerman. Diabetes mellitus and idiopathic hemochromatosis. Amer. J. Med. Sci. 255:53–62, 1968.

41. Balcerzak, S. P., W. W. Peternel, and E. W. Heinle. Iron absorption in chronic pancreatitis. Gastroenterology 53:257–264, 1967.

42. Balcerzak, S. P., J. W. Vester, and A. P. Doyle. Effect of iron deficiency and red cell age on human erythrocyte catalase activity. J. Lab. Clin. Med. 67:742–756, 1966.

43. Balcerzak, S. P., M. P. Westerman, E. W. Heinle, and F. H. Taylor. Measurement of iron stores using deferoxamine. Ann. Intern. Med. 68:518–525, 1968.

44. Balcerzak, S. P., M. P. Westerman, R. E. Lee, and A. P. Doyle. Idiopathic haemochromatosis: A study of three families. Amer. J. Med. 40:857–873, 1966.

45. Bannerman, R. M., S. T. Callender, R. M. Hardisty, and R. S. Smith. Iron absorption in thalassaemia. Brit. J. Haematol. 10:490–495, 1964.

46. Barry, M., D. M. Flynn, E. A. Letsky, and R. A. Risdon. Long-term chelation therapy in thalassaemia major: Effect on liver iron concentration, liver histology, and clinical progress. Brit. Med. J. 2:16–20, 1974.

47. Barshad, I. Chemistry of soil development, pp. 1–70. In F. E. Bear, Ed. Chemistry of the Soil. (2nd ed.) American Chemical Society Monograph Series No. 160. New York: Reinhold Publishing Corp., 1964.

48. Beadle, G. W. Yellow stripe—a factor for chlorophyll deficiency in maize located in the *Pr pr* chromosome. Amer. Natur. 63:189–192, 1929.

49. Beal, V. A., A. J. Meyers, and R. W. McCammon. Iron intake, hemoglobin, and physical growth during the first two years of life. Pediatrics 30:518–539, 1962.

50. Beaton, G. H. Epidemiology of iron deficiency, pp. 477–528. In A. Jacobs and M. Worwood, Eds. Iron in Biochemistry and Medicine. New York: Academic Press, 1974.

51. Beaton, G. H., Myo Thein, H. Milne, and M. J. Veen. Iron requirements of menstruating women. Amer. J. Clin. Nutr. 23:275–283, 1970.

52. Becking, G. C. Influence of dietary iron levels on hepatic drug metabolism *in vivo* and *in vitro* in the rat. Biochem. Pharmacol. 21:1585–1593, 1972.

53. Beevers, L., and R. H. Hageman. Nitrate reduction in higher plants. Ann. Rev. Plant Physiol. 20:495–522, 1969.

54. Beischer, N. A., R. Sivasamboo, S. Vohra, S. Silpisornkosal, and S. Reid. Placental hypertrophy in severe pregnancy anaemia. J. Obstet. Gynaecol. Brit. Commonw. 77:398–409, 1970.

55. Bell, W. D., L. Bogorad, and W. J. McIlrath. Response on the yellow-stripe maize mutant (ys_1) to ferrous and ferric iron. Bot. Gaz. 120:36–39, 1958.

56. Bender, M. L. Manganese nodules, pp. 673–677. In R. W. Fairbridge, Ed. The Encyclopedia of Geochemistry and Environmental Sciences. New York: Van Nostrand Reinhold Company, 1972.

57. Benjamin, B. I., S. Cortell, and M. E. Conrad. Bicarbonate-induced iron

complexes and iron absorption: One effect of pancreatic secretions. Gastroenterology 53:389–396, 1967.

58. Bernát, I. Ozaena: A Manifestation of Iron Deficiency. New York: Permagon Press, 1965, 116 pp.
59. Bernát, I., and J. Valló. Ozaena: The causes of its familial occurrence. Acta Med. Acad. Sci. Hung. 20:89–105, 1964.
60. Berner, R. Iron—abundance in natural waters, Section 26-I, 2 pp. In K. H. Wedepohl, Ed. Handbook of Geochemistry. Vol. II/2. New York: Springer-Verlag, 1970.
61. Berner, R. A. Principles of Chemical Sedimentology. New York: McGraw-Hill Book Company, 1971. 240 pp.
62. Berry, C. L., and W. C. Marshall. Iron distribution in the liver of patients with thalassaemia major. Lancet 1:1031–1033, 1967.
63. Berry, J. A., and H. M. Reisenauer. The influence of molybdenum on iron nutrition of tomato. Plant Soil 27:303—313, 1967.
64. Bessis, M., and J. Caroli. A comparative study of hemochromatosis by electron microscopy. Gastroenterology 37:538–549, 1959.
65. Beutler, E. Iron enzymes in iron deficiency. VI. Aconitase activity and citrate metabolism. J. Clin. Invest. 38:1605–1616, 1959.
66. Beutler, E., and R. K. Blaisdell. Iron enzymes in iron deficiency. III. Catalase in rat red cells and liver with some further observations on cytochrome c. J. Lab. Clin. Med. 52:694–699, 1958.
67. Beutler, E., and R. K. Blaisdell. Iron enzymes in iron deficiency. V. Succinic dehydrogenase in rat liver, kidney and heart. Blood 15:30–35, 1960.
68. Beutler, E., W. Drennan, and M. Block. The bone marrow and liver in iron-deficiency anemia: A histopathologic study of sections with special reference to stainable iron content. J. Lab. Clin. Med. 43:427–439, 1954.
69. Beutler, E., S. E. Larsh, and C. W. Gurney. Iron therapy in chronically fatigued, nonanemic women: A double-blind study. Ann. Intern. Med. 52:378–394, 1960.
70. Beveridge, B. R., R. M. Bannerman, J. M. Evanson, and L. J. Witts. Hypochromic anaemia: A retrospective study and follow-up of 378 inpatients. Q. J. Med. 34:145–166, 1965.
71. Bezwoda, W. R., T. H. Bothwell, L. A. van der Walt, S. Kronheim, and B. L. Pimstone. An investigation into gonadal dysfunction in patients with idiopathic haemochromatosis. Clin. Endocrinol. 6:379–385, 1977.
72. Bezwoda, W. R., P. B. Disler, S. R. Lynch, R. W. Charlton, J. D. Torrance, D. Derman, T. H. Bothwell, R. B. Walker, and F. Mayet. Patterns of food iron absorption in iron-deficient white and Indian subjects and in venesected haemochromatotic patients. Brit. J. Haematol. 33:425–436, 1976.
73. Biddulph, O. Translocation of minerals in plants, pp. 261–275. In E. Truog, Ed. Mineral Nutrition of Plants. Madison: University of Wisconsin Press, 1961.
74. Biesecker, J. E., and J. R. George. Stream Quality in Appalachia as Related to Coal-Mine Drainage, 1965. U.S. Geological Survey Circular 526. Washington, D.C.: U.S. Government Printing Office, 1966. 27 pp.
75. Björn-Rasmussen, E. Iron absorption from wheat bread: Influence of various amounts of bran. Nutr. Metab. 16:101–110, 1974.
76. Björn-Rasmussen, E., and L. Hallberg. Iron absorption from maize: Effect of ascorbic acid on iron absorption from maize supplemented with ferrous sulphate. Nutr. Metab. 16:94–100, 1974.

77. Björn-Rasmussen, E., L. Hallberg, B. Isaksson, and B. Arvidsson. Food iron absorption in man: Applications of the two-pool extrinsic tag method to measure heme and nonheme iron absorption from the whole diet. J. Clin. Invest. 53:247–255, 1974.

78. Björn-Rasmussen, E., L. Hallberg, and R. B. Walker. Food iron absorption in man. I. Isotopic exchange between food iron and inorganic iron salt added to food: Studies on maize, wheat, and eggs. Amer. J. Clin. Nutr. 25:317–323, 1972.

79. Blaxter, K. L., G. A. M. Sharman, and A. M. MacDonald. Iron-deficiency anaemia in calves. Brit. J. Nutr. 11:234–246, 1957.

80. Bomford, A., R. J. Walker, and R. Williams. Treatment of iron overload including results in a personal series of 85 patients with idiopathic haemochromatosis, pp. 324–331. In H. Kief, Ed. Iron Metabolism and Its Disorders. New York: American Elsevier Publishing Co. Inc., 1975.

81. Bomford, R. R., and C. P. Rhoads. Refractory anaemia. I. Clinical and pathological aspects. Q. J. Med. 10:175–234, 1941.

82. Bond, J., and T. T. Jones. Iron chelates of polyaminocarboxylic acids. Trans. Faraday Soc. 55:1310–1318, 1959.

83. Bothwell, T. H. Iron overload in the Bantu, pp. 362–375. In F. Gross, Ed. Iron Metabolism: An International Symposium Sponsored by CIBA, Aix-en-Provence, 1st-5th July, 1963. Berlin: Springer-Verlag, 1964.

84. Bothwell, T. H. The relationship of transfusional haemosiderosis to idiopathic haemochromatosis. South Afr. J. Clin. Sci. 4:53–70, 1953.

85. Bothwell, T. H., and T. Alper. The cardiac complications of haemochromatosis: Report of a case with a review of the literature. South Afr. J. Clin. Sci. 2:226–238, 1951.

86. Bothwell, T. H., and B. A. Bradlow. Siderosis in the Bantu: A combined histopathological and chemical study. Arch. Path. 70:279–292, 1960.

87. Bothwell, T. H., and R. W. Charlton. Dietary iron overload, pp. 221–229. In H. Kief, Ed. Iron Metabolism and Its Disorders. New York: American Elsevier Publishing Co. Inc., 1975.

88. Bothwell, T. H., I. Cohen, O. L. Abrahams, and S. M. Perold. A familial study in idiopathic hemochromatosis. Amer. J. Med. 27:730–738, 1959.

89. Bothwell, T. H., and C. A. Finch. Iron Metabolism. Boston: Little, Brown, and Company, 1962. 440 pp.

90. Bothwell, T. H., and C. Isaacson. Siderosis in the Bantu: A comparison of the incidence in males and females. Brit. Med. J. 1:522–524, 1962.

91. Bothwell, T. H., G. Pirzio-Biroli, and C. A. Finch. Iron absorption. I. Factors influencing absorption. J. Lab. Clin. Med. 51:24–36, 1958.

92. Bothwell, T. H., H. Seftel, P. Jacobs, J. D. Torrance, and N. Baumslag. Iron overload in Bantu subjects: Studies on the availability of iron in Bantu beer. Amer. J. Clin. Nutr. 14:47–51, 1964.

93. Bothwell, T. H., B. van Lingen, T. Alper, and M. L. du Preez. The cardiac complications of hemochromatosis: Report of a case including radio-iron studies and a note on etiology. Amer. Heart J. 43:333–340, 1952.

94. Bowen, H. J. M. Trace Elements in Biochemistry. New York: Academic Press, 1966. 241 pp.

95. Boyd, J. T., R. Doll, J. S. Faulds, and J. Leiper. Cancer of the lung in iron ore (haematite) miners. Brit. J. Ind. Med. 27:97–105, 1970.

96. Boyer, J. F., and V. E. Gleason. Coal and coal mine drainage. J. Water Pollut. Control Fed. 48:1284–1287, 1976.

97. Bradlow, B. A., J. A. Dunn, and J. Higginson. The effect of cirrhosis on iron storage. Amer. J. Path. 39:221–237, 1961.

98. Brain, J. D., P. A. Valberg, S. P. Sorokin, and W. C. Hinds. An iron oxide aerosol suitable for animal exposures. Environ. Res. 7:13–26, 1974.

99. Brain, M. C., and A. Herdan. Tissue iron stores in sideroblastic anaemia. Brit. J. Haematol. 11:107–115, 1965.

100. Bramer, H. C. Water Pollution Control in the Carbon and Alloy Steel Industries. (Prepared for the U.S. Environmental Protection Agency) EPA-600/2-76-193. Carnegie, Penn.: Datagraphics, Inc., 1976. 274 pp.

101. Braude, R., A. G. Chamberlain, M. Kotarbínska, and K. G. Mitchell. The metabolism of iron in piglets given labelled iron either orally or by injection. Brit. J. Nutr. 16:427–449, 1962.

102. Bremner, I., and A. C. Dalgarno. Iron metabolism in the veal calf. The availability of different iron compounds. Brit. J. Nutr. 29:229–243, 1973.

103. Bricaire, H., J. Guillon, J. Leprat, and E. Modigliani-Kourilsky. Le syndrome endocrinien des hémochromatoses idiopathiques. (A propos de 88 observations). Bull. Soc. Med. Hop. Paris 117:909–928, 1966.

104. Brief, R. S., R. S. Ajemian, and R. G. Confer. Iron pentacarbonyl: Its toxicity, detection, and potential for formation. Amer. Ind. Hyg. Assoc. J. 28:21–30, 1967.

105. Brief, R. S., J. W. Blanchard, R. A. Scala, and J. H. Blacker. Metal carbonyls in the petroleum industry. Arch. Environ. Health 23:373–384, 1971.

106. Briggs, G. A. Plume Rise. AEC Critical Review Series. TID-25075. Oak Ridge, Tenn.: U.S. Atomic Energy Commission, 1969. 81 pp.

107. Brink, B., P. Disler, S. Lynch, P. Jacobs, R. Charlton, and T. Bothwell. Patterns of iron storage in dietary iron overload and idiopathic hemochromatosis. J. Lab. Clin. Med. 88:725–731, 1976.

108. Brise, H., and L. Hallberg. A method for comparative studies on iron absorption in man using two radioiron isotopes. Acta Med. Scand. 171(Suppl.):7–22, 1962.

109. Brown, E., M. W. Skougstad, and M. J. Fishman. Techniques of Water-Resources Investigations of the United States Geological Survey. Book 5. Chapter A1. Methods for Collection and Analysis of Water Samples for Dissolved Minerals and Gases. Washington, D.C.: U.S. Government Printing Office, 1970. 160 pp.

110. Brown, E. B., Jr., R. Dubach, D. E. Smith, C. Reynafarje, and C. V. Moore. Studies in iron transportation and metabolism. X. Long-term iron overload in dogs. J. Lab. Clin. Med. 50:862–893, 1957.

111. Brown, E. B., Jr., D. E. Smith, R. Dubach, and C. V. Moore. Lethal iron overload in dogs. J. Lab. Clin. Med. 53:591–606, 1959.

112. Brown, J. C. An evaluation of bicarbonate induced iron chlorosis. Soil Sci. 89:246–247, 1960.

113. Brown, J. C. Iron chlorosis in plants. Adv. Agron. 13:329–369, 1961.

114. Brown, J. C. Competition between phosphates and the plant for Fe from Fe^{2+} ferrozine. Agron. J. 64:240–243, 1972.

115. Brown, J. C., and J. E. Ambler. Iron-stress response in tomato (*Lycopersicon esculentum*) 1. Sites of Fe reduction, absorption and transport. Physiol. Plant. 31:221–224, 1974.

116. Brown, J. C., and J. E. Ambler. "Reductants" released by roots of Fe-deficient soybeans. Agron. J. 65:311–314, 1973.

117. Brown, J. C., J. E. Ambler, R. L. Chaney, and C. D. Foy. Differential

responses of plant geotypes to micronutrients, pp. 389–418. In J. J. Mortvedt, P. M. Giordano, and W. L. Lindsay, Eds. Micronutrients in Agriculture. Proceedings of a Symposium, 1971. Madison, Wisc.: Soil Science Society of America, Inc., 1972.

118. Brown, J. C., and R. L. Chaney. Effect of iron on the transport of citrate into the xylem of soybean and tomatoes. Plant Physiol. 47:836–840, 1971.

119. Brown, J. C., R. L. Chaney, and J. E. Ambler. A new tomato mutant inefficient in the transport of iron. Physiol. Plant. 25:48–53, 1971.

120. Brown, J. C., and S. B. Hendricks. Enzymatic activities as indications of copper and iron deficiencies in plants. Plant Physiol. 27:651–660, 1952.

121. Brown, J. C., R. S. Holmes, and L. O. Tiffin. Iron chlorosis in soybeans as related to the genotype of rootstock. Soil Sci. 86:75–82, 1958.

122. Brown, J. C., R. S. Holmes, and L. O. Tiffin. Iron chlorosis in soybeans as related to the genotype of rootstalk: 3. Chlorosis susceptibility and reductive capacity at the root. Soil Sci. 91:127–132, 1961.

123. Brown, J. C., and W. E. Jones. A technique to determine iron efficiency in plants. Soil Sci. Soc. Amer. 40:398–405, 1976.

124. Brown, J. C., and W. E. Jones. Heavy-metal toxicity in plants. 1. A crisis in embryo. Commun. Soil Sci. Plant Anal. 6:421–438, 1975.

125. Brown, J. C., and W. E. Jones. pH changes associated with iron-stress response. Physiol. Plant. 30:148–152, 1974.

126. Brown, J. C., and W. E. Jones. Phosphorus efficiency as related to iron efficiency in sorghum. Agron. J. 67:468–472, 1975.

127. Brown, J. C., O. R. Lunt, R. H. Holmes, and L. O. Tiffin. The bicarbonate ion as an indirect cause of iron chlorosis. Soil Sci. 88:260–266, 1959.

128. Brown, J. C., and L. O. Tiffin. Iron stress as related to the iron and citrate occurring in stem exudate. Plant Physiol. 40:395–400, 1965.

129. Brown, J. C., L. O. Tiffin, and R. S. Holmes. Competition between chelating agents and roots as factor affecting absorption of iron and other ions by plant species. Plant Physiol. 35:878–886, 1960.

130. Brown, J. C., L. O. Tiffin, A. W. Specht, and J. W. Resnicky. Iron absorption by roots as affected by plant species and concentration of chelating agent. Agron. J. 53:81–85, 1961.

131. Brumfitt, W. Primary iron-deficiency anaemia in young men. Q. J. Med. 29:1–18, 1960.

132. Bryant, P. M. Methods of Estimation of the Dispersion of Windborne Material and Data to Assist in Their Application. Report AHSB(RP)-R-42. Harwell, Didcot, Berkshire: United Kingdom Atomic Energy Authority, 1964. 36 pp.

133. Bryson, R. A., and D. A. Baerreis. Possibilities of major climatic modifications and their implications: Northwest India, a case for study. Bull. Amer. Meteorol. Soc. 48:136–142, 1967.

134. Buchanan, W. M. Bantu siderosis.—A review. Central Afr. J. Med. 15:105–113, 1969.

135. Budowsky, E. I. The mechanism of the mutagenic action of hydroxylamines. Prog. Nucleic Acid Res. Molec. Biol. 16:125–188, 1976.

136. Buja, L. M., and W. C. Roberts. Iron in the heart. Etiology and clinical significance. Amer. J. Med. 51:209–221, 1971.

137. Bullen, J. J., H. J. Rogers, and E. Griffiths. Bacterial iron metabolism in infection and immunity, pp. 518–551. In J. B. Neilands, Ed. Microbial Iron Metabolism: A Comprehensive Treatise. New York: Academic Press, 1974.

138. Bulman, R. A. A critique of the chemistry of plutonium and the transuranics in the biosphere. Struct. Bond. (in press)

139. Bunn, C. R., and G. Matrone. In vivo interactions of cadmium, copper, zinc and iron in the mouse and rat. J. Nutr. 90:395–399, 1966.

140. Burks, J. M., M. A. Siimes, W. C. Mentzer, and P. R. Dallman. Iron deficiency in an Eskimo village. The value of serum ferritin in assessing iron nutrition before and after a three-month period of iron supplementation. J. Pediatr. 88:224–228, 1976.

141. Burman, D. Haemoglobin levels in normal infants aged 3 to 24 months and the effect of iron. Arch. Dis. Child. 47:261–271, 1972.

142. Byers, B. R., and C. E. Lankford. Regulation of synthesis of 2,3-dihydroxybenzoic acid in *Bacillus subtilis* by iron and a biological secondary hydroxamate. Biochim. Biophys. Acta 165:563–566, 1968.

143. Callender, S., D. G. Grahame-Smith, H. F. Woods, and M. B. H. Youdim. Reduction of platelet monoamine oxidase activity in iron deficiency anaemia. Brit. J. Pharmacol. 52:447P–448P, 1974.

144. Callender, S. T. Iron deficiency due to malabsorption of food iron, pp. 168–175. In H. Kief, Ed. Iron Metabolism and Its Disorders. Proceedings of the Third Workshop Conference Hoechst, Schloss Reisensburg, 6–9 April, 1975. International Congress Series 366. New York: American Elsevier Publishing Co., Inc., 1975.

145. Callender, S. T., S. R. Marney, Jr., and G. T. Warner. Eggs and iron absorption. Brit. J. Haematol. 19:657–665, 1970.

146. Callender, S. T., L. J. Witts, G. T. Warner, and R. Oliver. The use of a simple whole-body counter for haematological investigations. Brit. J. Haematol. 12:276–282, 1966.

147. Cammock, E. E., A. Turnbull, C. L. Fausnaugh, F. J. Cleton, L. M. Nyhus, C. A. Finch, and H. N. Harkins. Iron absorption from food after subtotal gastrectomy in the dog. Surg. Forum 12:299–301, 1961.

148. Canale, V. C., P. Steinherz, M. New, and M. Erlandson. Endocrine function in thalassemia major. Ann. N.Y. Acad. Sci. 232:333–345, 1974.

149. Cann, H. M., and H. L. Verhulst. Accidental poisoning in young children. The hazards of iron medication. A.M.A. J. Dis. Child. 99:688–691, 1960.

150. Cappell, D. F., H. E. Hutchison, and M. Jowett. Transfusional siderosis: The effects of excessive iron deposits on the tissues. J. Path. Bacteriol. 74:245–264, 1957.

151. Capriles, L. F. Intracranial hypertension and iron-deficiency anemia: Report of four cases. Arch. Neurol. 9:147–153, 1963.

152. Card, R. T., and M. C. Brain. The "anemia" of childhood. Evidence for a physiologic response to hyperphosphatemia. New Engl. J. Med. 228:388–392, 1973.

153. Carlson, T. N., and J. M. Prospero. The large-scale movement of Saharan air outbreaks over the northern equatorial Atlantic. J. Appl. Meteorol. 11:283–297, 1972.

154. Carr, J. G., C. V. Cutting, and G. C. Whiting, Eds. Lactic Acid Bacteria in Beverages and Foods. Proceedings of a Symposium, University of Bristol, 1973. New York: Academic Press, 1975. 415 pp.

155. Carroll, D. Role of clay minerals in the transportation of iron. Geochim. Cosmochim. Acta 14:1–28, 1958.

156. Cavill, I., and C. Ricketts. The kinetics of iron metabolism, pp. 613–647. In A. Jacobs and M. Worwood, Eds. Iron in Biochemistry and Medicine. New York: Academic Press, 1974.

157. Cawse, P. A. A Survey of Atmospheric Trace Elements in the U. K. (1972–73). Report No. AERE-R-7669. Harwell: U. K. Atomic Energy Research Establishment, 1974. 96 pp.

158. Černelč, M., V. Kušar, and S. Jeretin. Fatal peroral iron poisoning in a young woman. Acta Haematol. 40:90–94, 1968.

159. Chaberek, S., and A. E. Martell. Metal chelates in biological systems, pp. 416–504. In Organic Sequestering Agents. New York: John Wiley and Sons, 1959.

160. Chandra, R. K., B. Au, G. Woodford, and P. Hyam. Iron status in human response and susceptibility to infection, pp. 249–262. In R. Porter and W. Fitzsimons, Eds. Iron Metabolism. Ciba Foundation Symposium No. 51 (new series). Amsterdam: Elsevier, 1977.

161. Chaney, R. L., J. C. Brown, and L. O. Tiffin. Obligatory reduction of ferric chelates in iron uptake by soybeans. Plant Physiol. 50:208–213, 1972.

162. Chang, L. L. Tissue storage iron in Singapore. Amer. J. Clin. Nutr. 26:952–957, 1973.

163. Charlton, R. W., C. Abrahams, and T. H. Bothwell. Idiopathic hemochromatosis in young subjects: Clinical, pathological, and chemical findings in four patients. Arch. Path. 83:132–140, 1967.

164. Charlton, R. W., and T. H. Bothwell. Hemochromatosis: Dietary and genetic aspects. Prog. Hematol. 5:298–323, 1966.

165. Charlton, R. W., T. H. Bothwell, F. G. H. Mayet, C. J. Uys, and I. W. Simson. Liver iron stores in different population groups in South Africa. South Afr. Med. J. 45:524–529, 1971.

166. Charlton, R. W., T. H. Bothwell, and H. C. Seftel. Dietary iron overload. Clin. Haematol. 2:383–403, 1973.

167. Charlton, R. W., D. M. Hawkins, W. O. Mavor, and T. H. Bothwell. Hepatic storage iron concentrations in different population groups. Amer. J. Clin. Nutr. 23:358–371, 1970.

168. Charlton, R. W., P. Jacobs, H. Seftel, and T. H. Bothwell. Effect of alcohol on iron absorption. Brit. Med. J. 2:1427–1429, 1964.

169. Cheney, B. A., K. Lothe, E. H. Morgan, S. K. Sood, and C. A. Finch. Internal iron exchange in the rat. Amer. J. Physiol. 212:376–380, 1967.

170. Chisholm, M. The association between webs, iron and post-cricoid carcinoma. Postgrad. Med. J. 50:215–219, 1974.

171. Chisholm, M., G. M. Ardran, S. T. Callender, and R. Wright. A follow-up study of patients with post-cricoid webs. Q. J. Med. 40:409–420, 1971.

172. Chisholm, M., G. M. Ardran, S. T. Callender, and R. Wright. Iron deficiency and autoimmunity in post-cricoid webs. Q. J. Med. 40:421–433, 1971.

173. Chisolm, J. J., Jr. Poisoning due to heavy metals. Pediatr. Clin. North Amer. 17:591–615, 1970.

174. Chvapil, M., J. Hurych, and E. Ehrlichova. The effect of iron deficiency on the synthesis of collagenous and non-collagenous proteins in wound granulation tissue and in the heart of rats. Exp. Med. Surg. 26:52–60, 1968.

175. Clark, R. B., and J. C. Brown. Internal root control of iron uptake and utilization in maize genotypes. Plant Soil 40:669–677, 1974.

176. Clark, R. B., L. O. Tiffin, and J. C. Brown. Organic acids and iron translocation in maize genotypes. Plant Physiol. 52:147–150, 1973.

177. Clarke, F. W., Ed. The Data of Geochemistry. (5th ed.) U.S. Geological Survey Bulletin 770. Washington, D.C.: U.S. Government Printing Office, 1924. 841 pp.

178. Cole, P., and M. B. Goldman. Occupation, pp. 167–184. In J. F. Fraumeni, Jr., Ed. Persons at High Risk of Cancer. An Approach to Cancer Etiology and Control. New York: Academic Press, Inc., 1975.

179. Cole, S. K., W. Z. Billewicz, and A. M. Thomson. Sources of variation in menstrual blood loss. J. Obstet. Gynaecol. Brit. Commonw. 78:933–939, 1971.

180. Cole, S. K., A. M. Thomson, W. Z. Billewicz, and A. E. Black. Haematological characteristics and menstrual blood losses. J. Obstet. Gynaecol. Brit. Commonw. 79:994–1001, 1972.

181. Community Air Quality Guides. Iron oxide. Amer. Ind. Hyg. Assoc. J. 29:1–6, 1968.

182. Connor, J. J., and H. T. Shacklette. Background Geochemistry of Some Rocks, Soils, Plants, and Vegetables in the Conterminous United States. U.S. Geological Survey Professional Paper 574-F. Washington, D.C.: U.S. Government Printing Office, 1975. 168 pp.

183. Conrad, M. E. Factors affecting iron absorption, pp. 87–120. In L. Hallberg, H.-G. Harwerth, and A. Vannotti, Eds. Iron Deficiency. Pathogenesis, Clinical Aspects, Therapy. New York: Academic Press, 1970.

184. Conrad, M. E., B. I. Benjamin, H. L. Williams, and A. L. Foy. Human absorption of hemoglobin-iron. Gastroenterology 53:5–10, 1967.

185. Conrad, M. E., L. R. Weintraub, and W. H. Crosby. Iron metabolism in rats with phenylhydrazine-induced hemolytic disease. Blood 25:990–998, 1965.

186. Conrad, M. E., L. R. Weintraub, and W. H. Crosby. The role of the intestine in iron kinetics. J. Clin. Invest. 43:963–974, 1964.

187. Conrad, M. E., Jr., and W. H. Crosby. Intestinal mucosal mechanisms controlling iron absorption. Blood 22:406–415, 1963.

188. Cook, J. D. An evaluation of adsorption methods for measurement of plasma iron-binding capacity. J. Lab. Clin. Med. 76:497–506, 1970.

189. Cook, J. D., J. Alvarado, A. Gutnisky, M. Jamra, J. Labardini, M. Layrisse, J. Linares, A. Loría, V. Maspes, A. Restrepo, C. Reynafarje, L. Sánchez-Medal, H. Vélez, and F. Viteri. Nutritional deficiency and anemia in Latin America: A collaborative study. Blood 38:591–603, 1971.

190. Cook, J. D., W. E. Barry, C. Hershko, G. Fillet, and C. A. Finch. Iron kinetics with emphasis on iron overload. Amer. J. Path. 72:337–343, 1973.

191. Cook, J. D., G. M. Brown, and L. S. Valberg. The effect of achylia gastrica on iron absorption. J. Clin. Invest. 43:1185–1191, 1964.

192. Cook, J. D., C. A. Finch, and N. J. Smith. Evaluation of the iron status of a population. Blood 48:449–455, 1976.

193. Cook, J. D., C. Hershko, and C. A. Finch. Storage iron kinetics. V. Iron exchange in the rat. Brit. J. Haematol. 25:695–706, 1973.

194. Cook, J. D., M. Layrisse, and C. A. Finch. The measurement of iron absorption. Blood 33:421–429, 1969.

195. Cook, J. D., M. Layrisse, C. Martínez-Torres, R. Walker, E. Monsen, and C. A. Finch. Food iron absorption measured by an extrinsic tag. J. Clin. Invest. 51:805–815, 1972.

196. Cook, J. D., D. A. Lipschitz, L. E. M. Miles, and C. A. Finch. Serum ferritin as a measure of iron stores in normal subjects. Amer. J. Clin. Nutr. 27:681–687, 1974.

197. Cook, J. D., G. Marsaglia, J. W. Eschbach, D. D. Funk, and C. A. Finch. Ferrokinetics: A biologic model for plasma iron exchange in man. J. Clin. Invest. 49:197–205, 1970.

198. Cook, J. D., V. Minnich, C. V. Moore, A. Rasmussen, W. B. Bradley, and C. A. Finch. Absorption of fortification iron in bread. Amer. J. Clin. Nutr. 26:861–872, 1973.

199. Cook, J. D., and E. R. Monsen. Food iron absorption in human subjects. V. The effect of egg albumin and its amino acid components. Amer. J. Clin. Nutr. (in press)

200. Cook, J. D., and E. R. Monsen. Food iron absorption in man. II. The effect of EDTA on absorption of dietary non-heme iron. Amer. J. Clin. Nutr. 29:614–620, 1976.

201. Corn, M. Aerosols and the primary air pollutants—nonviable particles. Their occurrence, properties, and effects, pp. 77–168. In A. C. Stern, Ed. Air Pollution. (3rd ed.) Vol. 1. New York: Academic Press, 1976.

202. Cotes, J. E., J. M. Dabbs, P. C. Elwood, A. M. Hall, A. McDonald, and M. J. Saunders. The response to submaximal exercise in adult females; relation to haemoglobin concentration. J. Physiol. (Lond.) 203:79P–80P, 1969.

203. Council of Foods and Nutrition. Committee on Iron Deficiency. Iron deficiency in the United States. J.A.M.A. 203:407–412, 1968.

204. Cowan, J. W., M. Esfahani, J. P. Salji, and S. A. Azzam. Effect of phytate on iron absorption in the rat. J. Nutr. 90:423–427, 1966.

205. Creasia, D. A., and P. Nettesheim. Respiratory cocarcinogenesis studies with ferric oxide: A text case of current experimental models, pp. 234–245. In E. Karbe and J. F. Park, Eds. Experimental Lung Cancer. Carcinogenesis and Bioassays. New York: Springer-Verlag, 1974.

206. Crichton, R. R., H. Huebers, E. Huebers, D. Collet-Cassart, and Y. Ponce. Comparative studies on ferritin, pp. 193–200. In R. R. Crichton, Ed. Proteins of Iron Storage and Transport in Biochemistry and Medicine. New York: American Elsevier Publishing Company, Inc., 1975.

207. Cronan, D. S. Authigenic minerals in deep-sea sediments, pp. 491–525. In E. D. Goldberg, Ed. The Sea. Vol. 5. Marine Chemistry. New York: Wiley-Interscience, 1974.

208. Crosby, W. H. Serum ferritin fails to indicate hemochromatosis—nothing gold can stay. New Engl. J. Med. 294:333–334, 1976. (editorial)

209. Crosby, W. H., J. I. Munn, and F. W. Furth. Standardizing a method for clinical hemoglobinometry. U.S. Armed Forces Med. J. 5:693–703, 1954.

210. Crotty, J. J. Acute iron poisoning in children. Clin. Toxicol. 4:615–619, 1971.

211. Curtis, C. D., and D. A. Spears. The formation of sedimentary iron minerals. Econ. Geol. 63:257–270, 1968.

212. Dacie, J. V., and J. C. White. Erythropoesis with particular reference to its study by biopsy of human bone marrow: A review. J. Clin. Path. 2:1–32, 1949.

213. Dagg, J. H., A. Goldberg, J. R. Anderson, J. S. Beck, and K. G. Gray. Autoimmunity in iron-deficiency anaemia. Brit. Med. J. 1:1349–1350, 1964.

214. Dagg, J. H., J. M. Jackson, B. Curry, and A. Goldberg. Cytochrome oxidase in latent iron deficiency (sideropenia). Brit. J. Haematol. 12:331–333, 1966.

215. Dallman, P. R. Iron restriction in the nursing rat: Early effects upon tissue heme proteins, hemoglobin and liver iron. J. Nutr. 97:475–480, 1969.

216. Dallman, P. R. Tissue effects of iron deficiency, pp. 437–375. In A. Jacobs and M. Worwood, Eds. Iron and Biochemistry in Medicine. New York: Academic Press, 1974.

217. Dallman, P. R., and J. R. Goodman. Enlargement of the mitochondrial compartment in iron and copper deficiency. Blood 35:496–505, 1970.

218. Dallman, P. R., and J. R. Goodman. The effects of iron deficiency on the hepatocyte: A biochemical and ultrastructural study. J. Cell Biol. 48:79–90, 1971.

219. Dallman, P. R., and H. C. Schwartz. Distribution of cytochrome c and myoglobin in rats with dietary iron deficiency. Pediatrics 35:677–686, 1965.

220. Dallman, P. R., and H. C. Schwartz. Myoglobin and cytochrome response during repair of iron deficiency in the rat. J. Clin. Invest. 44:1631–1638, 1965.

221. Dallman, P. R., M. A. Siimes, and E. C. Manies. Brain iron: Persistent deficiency following short-term iron deprivation in the young rat. Brit. J. Haematol. 31:209–215, 1975.

222. Dallman, P. R., P. Sunshine, and Y. Leonard. Intestinal cytochrome response with repair of iron deficiency. Pediatrics 39:863–870, 1967.

223. David, C. N. Ferritin and iron metabolism in *Phycomyces*, pp. 149–158. In J. B. Neilands, Ed. Microbial Iron Metabolism: A Comprehensive Treatise. New York: Academic Press, 1974.

224. Davidson, W. M. B., and J. L. Markson. The gastric mucosa in iron deficiency anaemia. Lancet 2:639–643, 1955.

225. Davies, C. T. M., A. C. Chukweumeka, and J. P. M. van Haaren. Iron-deficiency anaemia: Its effect on maximum aerobic power and responses to exercise in African males aged 17–40 years. Clin. Sci. Mol. Med. 44:555–562, 1973.

226. Davis, P. N., L. C. Norris, and F. H. Kratzer. Iron deficiency studies in chicks. J. Nutr. 84:93–94, 1964. (letter)

227. Davis, P. N., L. C. Norris, and F. H. Kratzer. Iron utilization and metabolism in the chick. J. Nutr. 94:407–417, 1968.

228. Davison, R. H. Inheritance of haemochromatosis: A report on a family with consanguinity. Brit. Med. J. 2:1262–1263, 1961.

229. Debré, R., J-C. Dreyfus, J. Frézal, D. Labie, M. Lamy, P. Maroteaux, F. Schapira, and G. Schapira. Genetics of haemochromatosis. Ann. Hum. Genet. 23:16–30, 1958.

230. Deichmann, W. B., and W. H. Gerarde. Toxicology of Drugs and Chemicals. New York: Academic Press, 1969, 805 pp.

231. De Kimpe, C. R., and Y. A. Martel. Effects of vegetation on the distribution of carbon, iron, and aluminum in the B horizons of Northern Appalachian Spodosols. Soil Sci. Soc. Amer. J. 40:77–80, 1976.

232. DeKock, P. C. Fundamental aspects of iron nutrition of plants, pp. 41–44. In Ministry of Agriculture, Fisheries and Food, Technical Bulletin 21. Trace Elements in Soils and Crops. Proceedings of a Conference, 1966. London: Her Majesty's Stationery Office, 1971.

233. DeKock, P. C., and R. I. Morrison. The metabolism of chlorotic leaves. 2. Organic acids. Biochem. J. 70:272–277, 1958.

234. Delbarre, F. L'ostéoporose des hémochromatoses. Sem. Hop. Paris 36:3279–3294, 1960.

235. de Leeuw, N. K. M., L. Lowenstein, and Y-S. Hsieh. Iron deficiency and hydremia in normal pregnancy. Medicine (Baltimore) 45:291–315, 1966.

236. Deller, D. J. Iron[59] absorption measurements by whole-body counting: Studies in alcoholic cirrhosis, hemochromatosis, and pancreatitis. Amer. J. Dig. Dis. 10:249–258, 1965.

237. deMooy, C. J. Iron Deficiency in Soybeans—What Can Be Done About It? Pm-531 Cooperative Extension Service. Ames: Iowa State University, 1972. 2 pp.

238. de Sèze, S., J. Solnica, D. Mitrovic, L. Miravet, and H. Dartman. Joint and bone disorders and hypoparathyrodism in hemochromatosis. Semin. Arthritis Rheum. 2:71–94, 1972.

239. Desy, D. H. Iron and steel scrap, pp. 699–711. In U.S. Bureau of Mines. Minerals Yearbook 1974. Vol. 1. Metals, Minerals, and Fuels. Washington, D.C.: U.S. Government Printing Office, 1976.

240. de Villiers, J. M. Pedosesquioxides—composition and colloidal interactions in soil genesis during the Quaternary. Soil Sci. 107:454–461, 1969.

241. Diekmann, H. Siderochromes [iron(III)-trihydroxamates], pp. 449–457. In A. I. Laskin and H. A. Lechvalier, Eds. CRC Handbook of Microbiology. (Vol. 3) Cleveland: CRC Press, 1973.

242. Diez-Ewald, M., L. R. Weintraub, and W. H. Crosby. Interrelationship of iron and manganese metabolism. Proc. Soc. Exp. Biol. Med. 129:448–451, 1969.

243. Disler, P. B., S. R. Lynch, R. W. Charlton, J. D. Torrance, T. H. Bothwell, R. B. Walker, and F. Mayet. The effect of tea on iron absorption. Gut 16:193–200, 1975.

244. Domellof, L. Effects of Parenteral Iron Overload on the Rat Liver. Göteborg: Elanders Boktryckeri Aktieboleg, 1972. 83 pp.

245. Drysdale, J. W., P. Arosio, R. Adelman, J. T. Hazard, and D. Brooks. Isoferritins in normal and diseased states, pp. 359–366. In R. R. Crichton, Ed. Proteins of Iron Storage and Transport in Biochemistry and Medicine. New York: American Elsevier Publishing Company, Inc., 1975.

246. Dubach, R., C. V. Moore, and S. Callender. Studies in iron transportation and metabolism. IX. The excretion of iron as measured by the isotope technique. J. Lab. Clin. Med. 45:599–615, 1955.

247. Dubin, I. N. Idiopathic hemochromatosis and transfusion siderosis: A review. Amer. J. Clin. Path. 25:514–542, 1955.

248. Duce, R. A., P. L. Parker, (Eds.), and C. S. Giam. Pollutant Transfer to the Marine Environment. Deliberations and Recommendations of NSF/IDOE (National Science Foundation, International Decade of Ocean Exploration) Pollutant Transfer Workshop, held in Port Aransas, Texas, Jan. 11–12, 1974. Kingston: University of Rhode Island Press, 1974. 55 pp.

249. du Lac, Y., G. Deloux, and R. Deuil. Arthropathies et chondrocalcinoses au cours des hémochromatoses. Rev. Rhum. 34:758–769, 1967. (summary in English)

250. Dunn, W. L. Iron-loading, fibrosis, and hepatic carcinogenesis. Arch. Path. 83:258–266, 1967.

251. Dymock, I. W., J. Cassar, D. A. Pyke, W. G. Oakley, and R. Williams. Observations on the pathogenesis, complications and treatment of diabetes in 115 cases of haemochromatosis. Amer. J. Med. 52:203–210, 1972.

252. Dymock, I. W., E. B. D. Hamilton, J. W. Laws, and R. Williams. Arthropathy of hemochromatosis: Clinical and radiological analysis of 63 patients with iron overload. Ann. Rheum. Dis. 29:469–476, 1970.

253. Eakins, J. D., and D. A. Brown. An improved method for the simultaneous determination of iron-55 and iron-59 in blood by liquid scintillation counting. Int. J. Appl. Radiat. Isot. 17:391–397, 1966.

254. Easley, R. M., Jr., B. F. Schriener, Jr., and P. N. Yu. Reversible cardiomyopathy associated with hemochromatosis. New Engl. J. Med. 287:866–867, 1972.

255. Edgerton, V. R., S. L. Bryant, C. A. Gillespie, and G. W. Gardner. Iron deficiency anemia and physical performance and activity of rats. J. Nutr. 102:381–399, 1972.

256. Ediger, R. D., G. E. Peterson, and J. D. Kerber. Application of the graphite furnace to saline water analysis. Atom. Absorpt. Newslett. 13:61–64, 1974.

257. Eisenbud, M. The primary air pollutants—radioactive. Their occurrence, sources, and effects, pp. 197–231. In A. C. Stern, Ed. Air Pollution. (3rd ed.) Vol. 1. New York: Academic Press, 1976.

258. Ekenved, G. Iron absorption studies: Studies on oral iron preparations using serum iron and different radioiron isotope techniques. Scand. J. Haematol. (Suppl. 28):1–30, 1976.

259. Ekenved, G., L. Halvorsen, and L. Sölvell. Influence of a liquid antacid on the absorption of different iron salts. Scand. J. Haematol. (Suppl. 28):65–77, 1976.

260. Ellis, B. G., and B. D. Knezek. Adsorption reactions of micronutrients in soils, pp. 59–78. In J. J. Mortvedt, P. M. Giordano, and W. L. Lindsay, Eds. Micronutrients in Agriculture. Madison, Wis.: Soil Science Society of America, 1972.

261. El Wakeel, S. K., and J. P. Riley. Chemical and mineralogical studies of deep-sea sediments. Geochim. Cosmochim. Acta 25:110–146, 1961.

262. Elwood, P. C., and D. Hughes. Clinical trial of iron therapy on psychomotor function in anaemic women. Brit. Med. J. 3:254–255, 1970.

263. Elwood, P. C., A. Jacobs, R. G. Pitmin, and C. C. Entwistle. Epidemiology of the Paterson-Kelly syndrome. Lancet 2:716–720, 1964.

264. Elwood, P. C., W. E. Waters, W. J. W. Greene, P. Sweetnam, and M. M. Wood. Symptoms and circulating haemoglobin level. J. Chron. Dis. 21:615–628, 1969.

265. Emery, T. Biosynthesis and mechanism of action of hydroxamate-type siderochromes, pp. 107–123. In J. B. Neilands, Ed. Microbial Iron Metabolism: A Comprehensive Treatise. New York: Academic Press, 1974.

266. Engle, M. E., I. G. Erlandson, and C. H. Smith. Late cardiac complications of chronic, severe, refractory anemia with hemochromatosis. Circulation 30:698–705, 1964.

267. Epstein, J. H., and A. G. Redeker. Porphyria cutanea tarda: A study of the effect of phlebotomy. New Engl. J. Med. 279:1301–1304, 1968.

268. Eriksson, F., S. V. Johansson, H. Mellstedt, O. Strandberg, and P. O. Wester. Iron intoxication in two adult patients. Acta Med. Scand. 196:231–236, 1974.

269. Erlandson, M. E., B. Walden, G. Stern, M. W. Hilgartner, J. Wehman, and C. H. Smith. Studies on congenital hemolytic syndromes. IV. Gastrointestinal absorption of iron. Blood 19:359–378, 1962.

270. Eugster, H. P., and I.-M. Chou. The depositional environments of Precambrian banded iron-formations. Econ. Geol. 68:1144–1168, 1973.

271. Evans, H. J. Role of molybdenum in plant nutrition. Soil Sci. 81:199–208, 1956.

272. Evans, H. J. The biochemical role of iron in plant metabolism, pp. 89–110. In Mineral Nutrition of Trees. A Symposium. Duke University School of Forestry Bulletin 15. Durham, N.C.: Duke University, 1959.

273. Evans, J. Treatment of heart failure in haemochromatosis. Brit. Med. J. 1:1075–1078, 1959.

274. Eyster, E. Traumatic hemolysis with hemoglobinuria due to ball variance. Blood 33:391–395, 1969.

275. Fairbanks, V. F., J. L. Fahey, and E. Beutler. Acute iron poisoning, pp. 359–374. In Clinical Disorders of Iron Metabolism. (2nd ed.) New York: Grune & Stratton, 1971.

276. Fairbanks, V. F., J. L. Fahey, and E. Beutler. Clinical Disorders of Iron Metabolism. (2nd ed.) New York: Grune & Stratton, 1971. 486 pp.

277. Faulds, J. S. Haematite pneumoconiosis in Cumberland miners. J. Clin. Path. 10:187–199, Aug. 1957.

278. Faulds, J. S., and M. J. Stewart. Carcinoma of the lung in haematite miners. J. Path. Bacteriol. 72:353–366, 1956.

279. Fillet, G., J. D. Cook, and C. A. Finch. Storage iron kinetics. VII. A biologic model for reticuloendothelial iron transport. J. Clin. Invest. 53:1527–1533, 1974.

280. Fillet, G., and G. Marsaglia. Idiopathic hemochromatosis (IH). Abnormality in RBC transport of iron by the reticuloendothelial system (RES). Blood 46:1007, 1975. (abstract)

281. Finch, C. A. Body iron exchange in man. J. Clin. Invest. 38:392–396, 1959.

282. Finch, C. A. Iron metabolism in hemochromatosis. J. Clin. Invest. 28:780–781, 1949.

283. Finch, C. A., K. Deubelbeiss, J. D. Cook, J. W. Eschbach, L. A. Harker, D. D. Funk, G. Marsaglia, R. S. Hillman, S. Slichter, J. W. Adamson, A. Ganzoni, and E. R. Giblett. Ferrokinetics in man. Medicine (Baltimore) 49:17–53, 1970.

284. Finch, C. A., F. Hosain, E. H. Morgan, G. Marsaglia, E. Giblett, and R. S. Hillman. The ferrokinetic approach to anemia. Series Haematol. 6:30–40, 1965.

285. Finch, C. A., L. R. Miller, A. R. Inamdar, R. Person, K. Seiler, and B. Mackler. Iron deficiency in the rat: Physiological and biochemical studies of muscle dysfunction. J. Clin. Invest. 58:447–453, 1976.

286. Finch, S. C., and C. A. Finch. Idiopathic hemochromatosis, an iron storage disease. A. Iron metabolism in hemochromatosis. Medicine (Baltimore) 34:381–430, 1955.

287. Fischer, D. S., R. Parkman, and S. C. Finch. Acute iron poisoning in children: The problem of appropriate therapy. J.A.M.A. 218:1179–1184, 1971.

288. Fisher, M. W., P. E. Morrow, and C. L. Yuile. Effect of Freund's complete adjuvant upon the clearance of iron-59 oxide from rat lungs. J. Reticuloendothel. Soc. 13:536–556, 1973.

289. Forbes, G. Poisoning with a preparation of iron, copper, and manganese. Brit. Med. J. 1:367–370, 1947.

290. Ford, H. W. Sludges and Associated Problems Involving Agricultural Drains in Florida Wetlands. Paper Presented at 1970 Specialty Conference, American Society of Agricultural Engineers and American Society of Civil Engineers, Miami, Florida, Nov. 4–6, 1970.

291. Forth, W., and W. Rummel. Iron absorption. Physiol. Rev. 53:724–792, 1973.

292. Foucar, F. H., B. S. Gordon, and S. Kaye. Death following ingestion of ferrous sulfate. Amer. J. Clin. Path. 18:971–973, 1948.

293. Foy, H., A. Kondi, and W. H. Austin. Effect of dietary phytate on fecal absorption of radioactive ferric chloride. Nature 183:691–692, 1959.

294. Frenkiel, F. N., and R. E. Munn, Eds. Turbulent Diffusion in Environmental Pollution. Proceedings of a Symposium held at Charlottesville, Virginia, April 8–14, 1973. Advances in Geophysics Vol. 18B. New York: Academic Press, 1974. 389 pp.

295. Frick, P. G. Iron deficiency in blood donors, pp. 511–517. In L. Hallberg, H.-G. Harwerth, and A. Vannotti, Eds. Iron Deficiency: Pathogenesis, Clinical Aspects, Therapy. New York: Academic Press, 1970.

296. Fritz, J. C., G. W. Pla, T. Roberts, J. W. Boehne, and E. L. Hove. Bio-

logical availability in animals of iron from common dietary sources. J. Agric. Food Chem. 18:647–651, 1970.

297. Frost, A. E., H. H. Freedman, S. J. Westerback, and A. E. Martell. Chelating tendencies of N,N'-ethylenebis-[2-(o-hydroxyphenyl)]-glycine. J. Amer. Chem. Soc. 80:530–536, 1958.

298. Fuhr, I., and H. Steenbock. The effect of dietary calcium, phosphorus, and vitamin D on the utilization of iron. I. The effect of phytic acid on the availability of iron. II. The effect of vitamin D on body iron and hemoglobin production. III. The relation of rickets to anemia. J. Biol. Chem. 147:59–75, 1943.

299. Fullerton, H. W. The iron-deficiency anaemia of late infancy. Arch. Dis. Child. 12:91–110, 1937.

300. Furugouri, K. Characteristic aspects of iron metabolism in piglets. Jap. Agric. Res. Q. (JARQ) 9:171–176, 1975.

301. Fuwa, K., W. E. C. Wacker, R. Druyan, A. F. Bartholomay, and B. L. Vallee. Nucleic acids and metals, II: Transition metals as determinants of the conformation of ribonucleic acids. Proc. Nat. Acad. Sci. 46:1298–1307, 1960.

302. Gaffuri, E. Conoscenze attuali sui tumori professionali dell'apparato respiratorio. Lotta Tuberc. 40:281–302, 1970.

303. Gale, E., J. Torrance, and T. Bothwell. The quantitative estimation of total iron stores in human bone marrow. J. Clin. Invest. 42:1076–1082, 1963.

304. Gandra, Y. R., and R. B. Bradfield. Energy expenditure and oxygen handling efficiency of anemic school children. Amer. J. Clin. Nutr. 24:1451–1456, 1971.

305. Garby, L., L. Irnell, and I. Werner. Iron deficiency in women of fertile age in a Swedish community. Acta Med. Scand. 185:113–117, 1969.

306. Garibaldi, J. A. Influence of temperature on the biosynthesis of iron transport compounds by *Salmonella typhimurium*. J. Bacteriol. 110:262–265, 1972.

307. Garibaldi, J. A., and J. B. Neilands. Formation of iron-binding compounds by micro-organisms. Nature 177:526–527, 1956.

308. Gardner, G. W., V. R. Edgerton, R. J. Barnard, and E. M. Bernauer. Cardiorespiratory, hematological and physical performance responses of anemic subjects to iron treatment. Amer. J. Clin. Nutr. 28:982–988, 1975.

309. Garrels, R. M., and C. L. Christ. Solutions, Minerals, and Equilibria. New York: Harper and Row, 1965. 450 pp.

310. Garrels, R. M., and F. T. Mackenzie. Evolution of Sedimentary Rocks. New York: W. W. Norton and Company, 1971. 397 pp.

311. Giblett, E. R. F. transferrin, pp. 555–568. In W. S. Root and N. I. Berlin, Eds. Physiological Pharmacology. Vol. V. Blood. New York: Academic Press, 1974.

312. Gitlow, S. E., and M. R. Beyers. Metabolism of iron. I. Intravenous iron tolerance tests in normal subjects and patients with hemochromatosis. J. Lab. Clin. Med. 39:337–346, 1952.

313. Glover, J., and A. Jacobs. Activity pattern of iron-deficient rats. Brit. Med. J. 2:627–628, 1972.

314. Golberg, L., L. E. Martin, and J. P. Smith. Iron overloading phenomena in animals. Toxicol. Appl. Pharmacol. 2:683–707, 1960.

315. Golberg, L., and J. P. Smith. Iron overloading and hepatic vulnerability. Amer. J. Path. 36:125–149, 1960.

316. Golberg, L., J. P. Smith, and L. E. Martin. The effects of intensive and prolonged administration of iron parenterally in animals. Brit. J. Exp. Path. 38:297–311, 1957.

317. Goldberg, E. C. Chemistry—the oceans as a chemical system, pp. 3–25. In M. N. Hill, Ed. The Sea. Vol. 2. Composition of Sea-Water, Comparative and Descriptive Oceanography. New York: Interscience Publishers, 1963.

318. Goodman, J. R., J. B. Warshaw, and P. R. Dallman. Cardiac hypertrophy in rats with iron and copper deficiency: Quantitative contribution of mitochondrial enlargement. Pediatr. Res. 4:244–256, 1970.

319. Goossens, J. P. Idiopathic haemochromatosis: Juvenile and familial type—endocrine aspects. Neth. J. Med. 18:161–169, 1975.

320. Goya, N., S. Miyazaki, S. Kodate, and B. Ushio. A family of congenital atransferrinemia. Blood 40:239–245, 1972.

321. Grace, N. D., and L. W. Powell. Iron storage disorders of the liver. Gastroenterology 67:1257–1283, 1974.

322. Grady, R. W., J. H. Graziano, H. A. Akers, and A. Cerami. The development of new iron-chelating drugs. J. Pharmacol. Exp. Ther. 196:478–495, 1976.

323. Green, R., R. Charlton, H. Seftel, T. Bothwell, F. Mayet, B. Adams, C. Finch, and M. Layrisse. Body iron excretion in man: A collaborative study. Amer. J. Med. 45:336–353, 1968.

324. Greenberg, G. Sarcoma after intramuscular iron injection. Brit. Med. J. 1:1508–1509, 1976.

325. Greenberg, M. S., G. Strohmeyer, G. J. Hine, W. H. Keene, G. Curtis, and T. C. Chalmers. Studies in iron absorption. III. Body radioactivity measurements of patients with liver disease. Gastroenterology 46:651–661, 1964.

326. Greengard, J., and J. T. McEnery. Iron poisoning in children. GP 37(2):88–93, 1968.

327. Greenwald, I. The effect of phosphate on the solubility of calcium carbonate and of bicarbonate on the solubility of calcium and magnesium phosphates. J. Biol. Chem. 161:697–704, 1945.

328. Greweling, T. The Chemical Analysis of Plant Tissue. Agronomy Mimeo 66-22. Ithaca, N.Y.: Cornell University, 1966. 82 pp.

329. Grosberg, S. J. Hemochromatosis and heart failure: Presentation of a case with survival after three years' treatment by repeated venesection. Ann. Intern. Med. 54:550–566, 1961.

330. Guha, D. K., B. N. Walia, B. N. Tandon, M. G. Deo, and O. P. Ghai. Small bowel changes in iron-deficiency anaemia of childhood. Arch. Dis. Child. 43:239–244, 1968.

331. Haematite and iron oxide, pp. 29–36. In IARC Monographs on the Evaluation of Carcinogenic Risk of Chemicals to Man. Vol. 1. Lyon, France: World Health Organization, 1972.

332. Hahn, P. F., E. Jones, R. C. Lowe, G. R. Meneely, and W. Peacock. The relative absorption and utilization of ferrous and ferric iron in anemia as determined with the radioactive isotope. Amer. J. Physiol. 143:191–197, 1945.

333. Halitsky, J. Gas diffusion near buildings, pp. 221–225. In D. H. Slade, Ed. Meteorology and Atomic Energy 1968. TID-24190. Oak Ridge, Tenn.: U.S. Atomic Energy Commission, Division of Technical Information, 1968.

334. Hall, D. O. Copper and iron metalloproteins. Trends Biochem. Sci. 2:N86, 1977. (letter)

335. Hall, D. O., R. Cammack, and K. K. Rao. Non-haem iron proteins, pp. 279–334. In A. Jacobs and M. Worwood, Eds. Iron in Biochemistry and Medicine. New York: Academic Press, 1974.

336. Hall, G. J. L., and A. E. Davis. Inhibition of iron absorption by magnesium trisilicate. Med. J. Austral. 2:95–96, 1969.

337. Hallberg, L., and E. Björn-Rasmussen. Determination of iron absorption from whole diet: A new two-pool model using two radioiron isotopes given as haem and non-haem iron. Scand. J. Haematol. 9:193–197, 1972.

338. Hallberg, L., A. -M. Högdahl, L. Nilsson, and G. Rybo. Menstrual blood loss—a population study. Variation at different ages and attempts to define normality. Acta Obstet. Gynecol. Scand. 45:320–351, 1966.

339. Hallberg, L., and L. Sölvell. Iron absorption studies. Acta Med. Scand. 168(Suppl. 358):1–108, 1960.

340. Hallberg, L., L. Sölvell, and B. Zederfeldt. Iron absorption after partial gastrectomy: A comparative study on the absorption from ferrous sulphate and hemoglobin. Acta Med. Scand. 179(Suppl. 445):269–275, 1966.

341. Hallgren, B., and P. Sourander. The effect of age on the non-haemin iron in the human brain. J. Neurochem. 3:41–51, 1958.

342. Hamilton, A., and H. L. Hardy. Industrial Toxicology. (3rd ed.) Acton, Mass.: Publishing Sciences Group, Inc., 1974. 575 pp.

343. Hamilton, A., and H. L. Hardy. Metal fumes, pp. 195–196; Benign dusts, pp. 197–201; and Mixed dusts, pp. 443–455. In Industrial Toxicology. (3rd ed.) Acton, Mass.: Publishing Sciences Group, Inc., 1974.

344. Hamilton, D. L., and L. S. Valberg. Relationship between cadmium and iron absorption. Amer. J. Physiol. 227:1033–1037, 1974.

345. Hamilton, E., R. Williams, K. A. Barlow, and P. M. Smith. The arthropathy of idiopathic haemochromatosis. Q. J. Med. 37:171–182, 1968.

346. Hankes, L. V., C. R. Jansen, and M. Schmaeler. Ascorbic acid catabolism in Bantu with hemosiderosis (scurvy). Biochem. Med. 9:244–255, 1974.

347. Hansen, H. A., and A. Weinfeld. Hemosiderin estimations and sideroblast counts in the differential diagnosis of iron deficiency and other anemias. Acta Med. Scand. 165:333–356, 1959.

348. Hård, S. Non-anemic iron deficiency as an etiologic factor in diffuse loss of hair of the scalp in women. Acta Derm. Venereol. 43:562–569, 1963.

349. Harker, L. A., D. D. Funk, and C. A. Finch. Evaluation of storage iron by chelates. Amer. J. Med. 45:105–115, 1968.

350. Harrison, P. M., R. J. Hoare, T. G. Hoy, and I. G. Macara. Ferritin and haemosiderin: Structure and function, pp. 73–114. In A. Jacobs and M. Worwood, Eds. Iron in Biochemistry and Medicine. New York: Academic Press, 1974.

351. Hartley, W. J., J. Mullins, and B. M. Lawson. Nutritional siderosis in the bovine. N. Z. Vet. J. 7:99–105, 1959.

352. Harvard University. Oxygenation of Ferrous Iron. Water Pollution Control Research Series 14010-06/69. Washington, D.C.: U.S. Government Printing Office, 1970. 201 pp.

353. Haskins, D., A. R. Stevens, Jr., S. Finch, and C. A. Finch. Iron metabolism. Iron stores in man as measured by phlebotomy. J. Clin. Invest. 31:543–547, 1952.

354. Hathorn, M. K. S. The influence of hypoxia on iron absorption in the rat. Gastroenterology 60:76–81, 1971.

355. Heck, W. W., and L. F. Bailey. Chelation of trace metals in nutrient solutions. Plant Physiol. 25:573–582, 1950.

356. Hegenauer, J., and P. Saltman. Iron and susceptibility to infectious disease. Science 188:1038–1039, 1975.

357. Hegsted, D. M., C. A. Finch, and T. D. Kinney. The influence of diet on iron absorption. II. The interrelation of iron and phosphorus. J. Exp. Med. 90:147–156, 1949.

358. Heinrich, H. C. Intestinal iron absorption in man—methods of measurement, dose relationship, diagnostic and therapeutic applications, pp. 213–296. In L. Hallberg, H.-G. Harwerth, and A. Vannotti, Eds. Iron Deficiency: Pathogenesis, Clinical Aspects, Therapy. London: Academic Press, 1970.

359. Hem, J. D. Chemical factors that influence the availability of iron and manganese in aqueous systems. Geol. Soc. Amer. Bull. 83:443–450, 1972.

360. Hem, J. D. Role of hydrous metal oxides in the transport of heavy metals in the environment (G. F. Lee). Prog. Water Technol. 7:149–153, 1975. (Discussion)

361. Hem., J. D. Study and Interpretation of the Chemical Characteristics of Natural Water. (2nd ed.) Geological Survey Water-Supply Paper 1473. Washington, D.C., 1970. 363 pp.

362. Hem, J. D., and W. H. Cropper. Survey of Ferrous-Ferric Chemical Equilibria and Redox Potentials. U.S. Geological Survey Water-Supply Paper 1459-A. Washington, D.C.: U.S. Government Printing Office, 1959. 31 pp.

363. Hemmaplardh, D., S. G. Kailis, and E. H. Morgan. The effects of inhibitors of microtubule and microfilament function on transferrin and iron uptake by rabbit reticulocytes and bone marrow. Brit. J. Haematol. 28:53–65, 1974.

364. Henry, M. C., and D. G. Kaufman. Clearance of benzo[a]pyrene from hamster lungs after administration on coated particles. J. Nat. Cancer Inst. 51:1961–1964, 1973.

365. Henry, M. C., C. D. Port, and D. G. Kaufman. Importance of physical properties of benzo(a)pyrene-ferric oxide mixtures in lung tumor induction. Cancer Res. 35:207–217, 1975.

366. Henry, M. C., C. D. Port, and D. G. Kaufman. Role of particles in respiratory carcinogenesis bioassay, pp. 173–185. In E. Karbe and J. F. Park, Eds. Experimental Lung Cancer: Carcinogenesis and Bioassays. International Symposium Held at the Battelle Seattle Research Center. Seattle, WA, USA, June 23–26, 1974. New York: Springer-Verlag, 1974.

367. Hershko, C., J. D. Cook, and C. A. Finch. Storage iron kinetics. II. The uptake of hemoglobin iron by hepatic parenchymal cells. J. Lab. Clin. Med. 80:624–634, 1972.

368. Hershko, C., J. D. Cook, and C. A. Finch. Storage iron kinetics. VI. The effect of inflammation on iron exchange in the rat. Brit. J. Haematol. 28:67–75, 1974.

369. Hershko, Ch., A. Karsai, L. Eylon, and G. Izak. The effect of chronic iron deficiency on some biochemical functions of the human hemopoietic tissue. Blood 36:321–329, 1970.

370. Heslop-Harrison, J. Development, differentiation and yield, pp. 291–321. In J. D. Eastin, F. A. Haskins, C. Y. Sullivan, and C. H. M. Van Bavel, Eds. Physiological Aspects of Crop Yield. Madison, Wisc.: American Society of Agronomy and Crop Science Society of America, 1969.

371. Hesse, P. R. A Textbook of Soil Chemical Analysis. New York: Chemical Publishing Co., 1971. 520 pp.

372. Hesseltine, C. W., C. Pidacks, A. R. Whitehill, N. Bohonos, B. L. Hutch-

ings, and J. H. Williams. Coprogen, a new growth factor for coprophilic fungi. J. Amer. Chem. Soc. 74:1326, 1952.

373. Hewitt, E. J. The essential nutrient elements: Requirements and interactions in plants, pp. 137–360. In F. C. Steward, Ed. Plant Physiology: A Treatise. Vol. 3. Inorganic Nutrition of Plants. New York: Academic Press, 1963.

374. Hewitt, E. J. Trace elements in plants: Biochemical aspects, pp. 21–34. In Ministry of Agriculture, Fisheries and Food, Technical Bulletin 21. Trace Elements in Soils and Crops. Proceedings of a Conference, 1966. London: Her Majesty's Stationery Office, 1971.

375. Hewitt, E. J., and S. C. Agarwala. Reduction of triphenyltetrazolium chloride by plant tissues and its relationship to molybdenum status. Nature 169:545–546, 1952.

376. Hidy, G. M., and J. R. Brock. An assessment of the global sources of tropospheric aerosols, pp. 1088–1097. In H. M. Englund and W. T. Beery, Eds. Proceedings of the Second International Clean Air Congress, Washington, D.C., 1970. New York: Academic Press, 1971.

377. Higginson, J., B. G. Grobbelaar, and A. R. P. Walker. Hepatic fibrosis and cirrhosis in man in relation to malnutrition. Amer. J. Path. 33:29–53, 1957.

378. Hill, C. H. Studies on the mutual interaction of iron and manganese on absorption from chick duodenal segments in situ. Fed. Proc. 30:236, 1971. (abstract)

379. Hill, C. H., and G. Matrone. Studies on copper and iron deficiencies in growing chickens, J. Nutr. 73:425–431, 1961.

380. Hillman, R. S., and C. A. Finch. Red Cell Manual. (4th ed.) Philadelphia: F. A. Davis Company, 1974. 84 pp.

381. Ho, W., and A. Furst. Intratracheal instillation method for mouse lungs. Oncology 27:385–393, 1973.

382. Hoag, M. S., R. O. Wallerstein, and M. Pollycove. Occult blood loss in iron deficiency anemia of infancy. Pediatrics 27:199–203, 1961.

383. Hobbs, J. R. Iron deficiency after partial gastrectomy. Gut 2:141–149, 1961.

384. Hoffbrand, A. V., and S. A. Broitman. Effect of chronic nutritional iron deficiency on the small intestinal disaccharidase activities of growing dogs. Proc. Soc. Exp. Biol. Med. 130:595–598, 1969.

385. Hoffbrand, A. V., K. Ganeshaguru, J. W. L. Hooton, and M. H. N. Tatersall. Effect of iron deficiency and desferrioxamine on DNA synthesis in human cells. Brit. J. Haematol. 33:517–526, 1976.

386. Hofvander, Y. Haematological investigations in Ethiopia with special reference to a high iron intake. Acta Med. Scand. Suppl. 494:1–74, 1968.

387. Hogan, G. R., and B. Jones. The relationship of koilonychia and iron deficiency in infants. J. Pediatr. 77:1054–1057, 1970.

388. Holmes, R. S., and J. C. Brown. Chelates as correctives for chlorosis. Soil Sci. 80:167–179, 1955.

389. Holowach, J., and D. L. Thurston. Breath-holding spells and anemia. New Engl. J. Med. 268:21–23, 1963.

390. Horowitz, N. H., G. Charlang, G. Horn, and N. P. Williams. Isolation and identification of the conidial germination factor of Neurospora crassa. J. Bacteriol. 127:135–140, 1976.

391. Howell, D. Significance of iron deficiencies. Consequences of mild deficiency in children, pp. 65–69. In National Research Council, Food and Nutrition Board. Extent and Meanings of Iron Deficiency in the U.S.

Proceedings of Workshop, March 8–9, 1971. Washington, D.C.: National Academy of Sciences, 1971.

392. Hübers, H., E. Hübers, W. Forth, G. Leopold, and W. Rummel. Binding of iron and other metals in brush borders of jejunem and ileum of the rat in vitro. Acta Pharmacol. Toxicol. 29(Suppl. 4):22, 1971.

393. Hübers, H., E. Hübers, W. Forth, and W. Rummel. Iron absorption and cellular transfer protein in the intestinal mucosa of mice with hereditary anemia. Naunyn-Schmiedeberg's Arch. Pharmacol. 274:R56, 1972. (abstract)

394. Hueper, W. C. A Quest into the Environmental Causes of Cancer of the Lung. Public Health Monograph No. 36. Washington, D.C.: U.S. Government Printing Office, 1955. 54 pp.

395. Hurtado, A., C. Merino, and E. Delgado. Influence of anoxemia on the hemopoietic activity. Arch. Intern. Med. 75:284–323, 1945.

396. Hussain, R., and V. N. Patwardhan. The influence of phytate on the absorption of iron. Indian J. Med. Res. 47:676–682, 1959.

397. Hutton, C. F. Plummer Vinson syndrome. Brit. J. Radiol. 29:81–85, 1956.

398. Hyde, B. B., A. J. Hodge, A. Kahn, and M. L. Birnstiel. Studies on phytoferrin. I. Identification and localization. J. Ultrastruct. Res. 9:248–258, 1963.

399. Hyman, C. B., B. Landing, R. Alfin-Slater, L. Kozak, J. Weitzman, and J. A. Ortega. dl-α-Tocopherol, iron, and lipofuscin in thalassemia. Ann. N.Y. Acad. Sci. 232:211–220, 1974.

400. Iljin, W. S. Metabolism of plants affected with lime-induced chlorosis (calciose). I. Nitrogen metabolism. Plant Soil 3:239–256, 1951.

401. Iljin, W. S. Metabolism of plants affected with lime-induced chlorosis (calciose). III. Mineral elements. Plant Soil 4:11–28, 1952.

402. International Committee for Standardization in Hematology. Recommended methods for radioisotope red cell survival studies. Blood 38:378–386, 1971.

403. Isaacson, C., H. C. Seftel, K. J. Keeley, and T. H. Bothwell. Siderosis in the Bantu: The relationship between iron overload and cirrhosis. J. Lab. Clin. Med. 58:845–853, 1961.

404. Ishinishi, N., Y. Kodama, E. Kunitake, K. Nobutomo, and Y. Fukushima. The carcinogenicity of dusts collected from an open-hearth furnace for the smelting of iron: A preliminary experimental study, pp. 480–488. In G. F. Nordberg, Ed. Effects and Dose-Response Relationships of Toxic Metals. New York: Elsevier Scientific Publishing Company, 1976.

405. Isied, S. S., G. Kuo, and K. N. Raymond. Coordination isomers of biological iron transport compounds. V. The preparation and chirality of the chromium(III) enterobactin complex and model tris (catechol)chromium(III) analogues. J. Amer. Chem. Soc. 98:1763–1767, 1976.

406. Ito, T., and J. B. Neilands. Products of "low-iron fermentation" with Bacillus subtilis: Isolation, characterization and synthesis of 2,3-dihydroxy-benzoylglycine. J. Amer. Chem. Soc. 80:4465–4647, 1958.

407. Iverson, W. P. Microbial corrosion of iron, pp. 475–513. In J. B. Neilands, Ed. Microbial Iron Metabolism: A Comprehensive Treatise. New York: Academic Press, 1974.

408. Jackson, M. L. Chemical composition of soils, pp. 71–141. In F. E. Bear, Ed. Chemistry of the Soil. (2nd ed.) American Chemical Society Monograph Series No. 160. New York: Reinhold Publishing Corp., 1964.

409. Jackson, T. L., J. Hay, and D. P. Moore. The effect of Zn on yield and chemical composition of sweet corn in the Willamette Valley. Amer. Soc. Hort. Sci. 91:462–471, 1967.

410. Jacobs, A. Anaemia and post-cricoid carcinoma. Brit. J. Cancer 15:736–744, 1961.

411. Jacobs, A. Iron-containing enzymes in the buccal epithelium. Lancet 2:1331–1333, 1961.

412. Jacobs, A. Iron deficiency: Effects on tissues and enzymes, pp. 135–141. In H. Kief, Ed. Iron Metabolism and Its Disorders. New York: American Elsevier Publishing Co. Inc., 1975.

413. Jacobs, A. Oral cornification in anaemic patients. J. Clin. Path. 12:235–237, 1959.

414. Jacobs, A. The buccal mucosa in anaemia. J. Clin. Path. 13:463–468, 1960.

415. Jacobs, A. Tissue changes in iron deficiency. Brit. J. Haematol. 16:1–4, 1969. (annotation)

416. Jacobs, A., and G. S. Kilpatrick. The Paterson-Kelly syndrome. Brit. Med. J. 2:79–82, 1964.

417. Jacobs, A., J. H. Lawrie, C. C. Entwistle, and H. Campbell. Gastric acid secretion in chronic iron-deficiency anaemia. Lancet 2:190–192, 1966.

418. Jacobs, A., and P. M. Miles. Role of gastric secretion in iron absorption. Gut 10:226–229, 1969.

419. Jacobs, A., F. Miller, M. Worwood, M. R. Beamish, and C. A. Wardrop. Ferritin in the serum of normal subjects and patients with iron deficiency and iron overload. Brit. Med. J. 4:206–208, 1972.

420. Jacobs, A., and M. Worwood. Ferritin in serum: Clinical and biochemical implications. New Engl. J. Med. 292:951–956, 1975.

421. Jacobs, P., T. Bothwell, and R. W. Charlton. Role of hydrochloric acid in iron absorption. J. Appl. Physiol. 19:187–188, 1964.

422. Jacobs, P., and C. A. Finch. Iron for erythropoiesis. Blood 37:220–230, 1971.

423. Jacobson, L. Maintenance of iron supply in nutrient solutions by a single addition of ferric potassium ethylenediamine tetraacetate. Plant Physiol. 26:411–413, 1951.

424. Jalili, M. A., and S. Al-Kassab. Koilonychia and cystine content of nails. Lancet 2:108–110, 1959.

425. James, H. L. Chemistry of the Iron-Rich Sedimentary Rocks. U.S. Geological Survey Professional Paper 440-W. Washington, D.C.: U.S. Government Printing Office, 1966. 61 pp.

426. Jandl, J. H., J. K. Inman, R. L. Simmons, and D. W. Allen, Transfer of iron from serum iron-binding protein to human reticulocytes. J. Clin. Invest. 38:161–185, 1959.

427. Jarvis, J. H., and A. Jacobs. Morphological abnormalities in lymphocyte mitochondria associated with iron-deficiency anaemia. J. Clin. Path. 27:973–979, 1974.

428. Jenkins, D. J. A., M. S. Hill, and J. H. Cummings. Effect of wheat fiber on blood lipids, fecal steroid excretion and serum iron. Amer. J. Clin. Nutr. 28:1408–1411, 1975.

429. Jenne, E. A. Controls on Mn, Fe, Co, Ni, Cu, and Zn concentrations in soils and water: The significant role of hydrous Mn and Fe oxides. Adv. Chem. Ser. 73:337–387, 1968.

430. Johnston, R. E., and W. M. Krogman. Patterns of growth in children with thalassemia major. Ann. N.Y. Acad. Sci. 119:667–679, 1964.

431. Jones, J. G., and C. G. Warner. Chronic exposure to iron oxide, chromium oxide, and nickel oxide fumes of metal dressers in a steel-works. Brit. J. Ind. Med. 29:169–177, 1972.

432. Joske, R. A., E. S. Finckh, and I. J. Wood. Gastric biopsy: A study of 1,000 consecutive successful gastric biopsies. Q. J. Med. 24:269–294, 1955.

433. Judisch, J. M., J. L. Naiman, and F. A. Oski. The fallacy of a fat iron-deficient child. Pediatrics 37:987–990, 1966.

434. Kailis, S. G., and E. H. Morgan. Transferrin and iron uptake by rabbit bone marrow cells *in vitro*. Brit. J. Haematol. 28:37–52, 1974.

435. Kattamis, C., N. Touliatos, S. Haidas, and N. Matsaniotis. Growth of children with thalassaemia: Effect of different transfusion regimens. Arch. Dis. Child. 45:502–505, 1970.

436. Kavin, H., R. W. Charlton, P. Jacobs, R. Green, J. D. Torrance, and T. H. Bothwell. Effect of the exocrine pancreatic secretions on iron absorption. Gut 8:556–564, 1967.

437. Kawai, M., T. Matsuyama, A. Sakai, and T. Yoshitsugu. Iron and steel, pp. 742–745. In Encyclopaedia of Occupational Safety and Health. Vol. 1. Geneva: International Labour Office, 1971.

438. Keller-Schierlein, W., V. Prelog, and H. Zähner. Siderochrome. (Natürliche Eisen(III)-trihydroxamat-Komplexe.) Fortschr. Chem. Org. Naturst. 22:279–322, 1964.

439. Kelly, A. B. Spasm of the entrance to the oesophagus. J. Laryngol. 34:285–289, 1919.

440. Kennedy, V. C., G. W. Zellweger, and B. F. Jones. Filter pore-size effects on the analysis of Al, Fe, Mn, and Ti in water. Water Resources Res. 10:785–790, 1974.

441. Kent, G., and H. Popper. Liver biopsy in diagnosis of hemochromatosis. Amer. J. Med. 44:837–841, 1968. (editorial)

442. Kent, G., F. I. Volini, O. T. Minick, E. Orfei, and J. de la Huerga. Effect of iron loading upon the formation of collagen in the hepatic injury induced by carbon tetrachloride. Amer. J. Path. 45:129–155, 1964.

443. Kim, Y. S., and D. F. Martin. On the effect of iron-citrate on the growth of the Florida red tide organism, *Gymnodinium breve*. Environ. Lett. 9:55–58, 1975.

444. Kimber, C., and L. R. Weintraub. Malabsorption of iron secondary to iron deficiency. New Engl. J. Med. 279:453–459, 1968.

445. Kincaid, R. L., and B. L. O'Dell. Enterochelin and unrecognized growth stimulants for guinea pigs. Fed. Proc. 33:666, 1974.

446. Kirkby, E. A., and K. Mengel. Ionic balance in different tissues of the tomato plant in relation to nitrate, urea, or ammonium nutrition. Plant Physiol. 42:6–14, 1967.

447. Klauder, D. S., and H. G. Petering. Protective value of dietary copper and iron against some toxic effects of lead in rats. Environ. Health Perspect. 12:77–80, 1975.

448. Kleckner, M. S., Jr., A. H. Baggenstoss, and J. F. Weir. Iron-storage diseases. Amer. J. Clin. Path. 25:915–931, 1955.

449. Klemic, H., H. L. James, and G. D. Eberlein. Iron, pp. 291–306. In D. A. Brobst and W. P. Pratt, Eds. United States Mineral Resources. (U.S. Geological Survey Professional Paper 820). Washington, D.C.: U.S. Government Printing Office, 1973.

450. Klinger, F. L. Iron ore. Preprint from Bureau of Mines Bulletin 667. A Chapter from Facts and Problems, 1975 Edition. Washington, D.C.: U.S. Government Printing Office, 1975. 21 pp.

451. Klinger, F. L. Iron ore, pp. 651–675. In Bureau of Mines. Minerals Yearbook 1974. Vol. 1. Metals, Minerals and Fuels. Washington, D.C.: U.S. Government Printing Office, 1976.

452. Knizley, H., Jr., and W. D. Noyes. Iron deficiency anemia, papill-edema, thrombocytosis, and transient hemiparesis. Arch. Intern Med. 129:483–486, 1972.

453. Kolb, E. The metabolism of iron in farm animals under normal and pathologic conditions. Adv. Vet. Sci. 8:49–114, 1963.

454. Kopp, J. F. Suggested method for spectrochemical analysis of water by the rotating disk technique using an optical emission spectrometer, pp. 1010–1015. In Methods for Emission Spectrochemical Analysis. (6th ed.) Philadelphia: American Society for Testing and Materials, 1971.

455. Krauskopf, K. B. Geochemistry of micronutrients, pp. 7–40. In J. J. Mortvedt, P. M. Giordano, and W. L. Lindsay, Eds. Micronutrients in Agriculture. Proceedings of a Symposium. Madison, Wisc.: Soil Science Society of America, Inc., 1972.

456. Kuhn, J. N., M. Layrisse, M. Roche, C. Martínez-Torres, and R. B. Walker. Observations on the mechanism of iron absorption. Amer. J. Clin. Nutr. 21:1184–1188, 1968.

457. Kuo, B., E. Zaino, and M. S. Roginsky. Endocrine function in thalassemia major. J. Clin. Endocrinol. Metab. 28:805–808, 1968.

458. Kurtz, H. F. Bauxite, pp. 211–223. In 1970 Bureau of Mines Minerals Yearbook. Vol. 1. Metals, Minerals, and Fuels. Washington, D.C.: U.S. Government Printing Office, 1972.

459. Kurtz, H. F. Bauxite, pp. 189–204. In 1972 Bureau of Mines Minerals Yearbook. Vol. 1. Metals, Minerals, and Fuels. Washington, D.C., U.S. Government Printing Office, 1974.

460. Kushner, J. P., G. R. Lee, and S. Nacht. The role of iron in the pathogenesis of porphyria cutanea tarda: An in vitro model. J. Clin. Invest. 51:3044–3011, 1972.

461. Lamont, N. McE., and M. Hathorn. Increased plasma iron and liver pathology in Africans with porphyria. South Afr. Med. J. 34:279–281, 1960.

462. Langer, E. E., R. G. Haining, R. F. Labbe, P. Jacobs, E. F. Crosby, and C. A. Finch. Erythrocyte protoporphyrin. Blood 40:112–128, 1972.

463. Langmuir, D. Particle size effect on the reaction goethite $\rightleftarrows$ hematite water. Amer. J. Sci. 271:147–156, 1971.

464. Langmuir, D. The Gibbs free energies of substances in the system $Fe-O_2-H_2O-CO_2$ at 25°C, pp. 180B–184B. In Geological Survey Research, 1969. (U.S. Geological Survey Professional Paper 650-B). Washington, D.C.: U.S. Government Printing Office, 1969.

465. Langmuir, D., and D. O. Whittemore. Variations in the stability of precipitated ferric oxyhydroxides. Adv. Chem. Ser. 106:209–234, 1971.

466. Lanzkowsky, P. The influence of maternal iron-deficiency anaemia on the hemoglobin of the infant. Arch. Dis. Child. 36:205–209, 1961.

467. Lassman, M. N., M. Genel, J. K. Wise, R. Hendler, and P. Felig. Carbohydrate homeostasis and pancreatic islet cell function in thalassemia. Ann. Intern. Med. 80:65–69, 1974.

468. Lassman, M. N., R. T. O'Brien, H. A. Pearson, J. K. Wise, R. K. Donabedian, P. Felig, and M. Genel. Endocrine evaluation in thalassemia major. Ann. N.Y. Acad. Sci. 232:226–237, 1974.

469. Latimer, W. M. Oxidation Potentials. The Oxidation States of Elements and Their Potentials in Aqueous Solutions. (2nd ed.) Englewood Cliffs, N.J.: Prentice-Hall, Inc., 1952. 392 pp.

470. Lavender, S., and J. A. Bell. Iron intoxication in an adult. Brit. Med. J. 2:406, 1970.

471. Lawlor, M. J., W. H. Smith, and W. M. Beeson, Iron requirement of the growing lamb. J. Anim. Sci. 24:742–747, 1965.

472. Layrisse, M. Dietary iron absorption, pp. 25–33. In H. Kief, Ed. Iron Metabolism and Its Disorders. Proceedings of the Third Workshop Conference Hoechst, Schloss Reisensburh, 6–9 April, 1975. International Congress Series 366. New York: American Elsevier Publishing Company, Inc., 1975.

473. Layrisse, M., J. D. Cooke, C. Martínez-Torres, M. Roche, I. N. Kuhn, R. B. Walker, and C. A. Finch. Food iron absorption: A comparison of vegetable and animal foods. Blood 33:430–443, 1969.

474. Layrisse, M., and C. Martínez-Torres. Model for measuring dietary absorption of heme iron: Test with a complete meal. Amer. J. Clin. Nutr. 25:401–411, 1972.

475. Layrisse, M., C. Martínez-Torres, J. D. Cook, R. Walker, and C. A. Finch. Iron fortification of food: Its measurement by the extrinsic tag method. Blood 41:333–352, 1973.

476. Layrisse, M., C. Martínez-Torres, and M. González. Measurement of the total daily dietary iron absorption by the extrinsic tag model. Amer. J. Clin. Nutr. 27:152–162, 1974.

477. Layrisse, M., C. Martínez-Torres, M. Renzy, and I. Leets. Ferritin iron absorption in man. Blood 45:689–698, 1975.

478. Layrisse, M., C. Martínez-Torres, and M. Roche. Effect of interaction of various foods on iron absorption. Amer. J. Clin. Nutr. 21:1175–1183, 1968.

479. Layrisse, M., A. Paz, N. Blumenfeld, and M. Roche. Hookworm anemia: Iron metabolism and erythrokinetics. Blood 18:61–72, 1961.

480. Layrisse, M., and M. Roche. The relationship between anemia and hookworm infection: Results of surveys of rural Venezuelan population. Amer. J. Hyg. 79:279–301, 1964.

481. Le Bouffant, L., J.-P. Henin, J.-C. Martin, and H. Da Niel. Etude expérimentale de l'epuration pulmonaire. Action de l'empoussiérage sur la capacite d'épuration. Application à l'homme. Lille Med. 17:1091–1101, 1972.

482. Lee, D. H. K. Biological effects of ingested asbestos: Report and commentary. Environ. Health Perspect. 9:113–122, 1975.

483. Lees, F., and F. D. Rosenthal. Gastric mucosal lesions before and after treatment in iron deficiency anaemia. Q. J. Med. 27:19–26, 1958.

484. Leonard, B. J. Gastric acid in iron-deficiency anaemia. Lancet 2:440–441, 1966.

485. Leonard, B. J. Hypochromic anaemia in R.A.F. recruits. Lancet 1:899–902, 1954.

486. Leonard, C. D., and I. Stewart. An available source of iron for plants. Proc. Amer. Soc. Hort. Sci. 62:103–110, 1953.

487. Leong, J., J. B. Neilands, and K. N. Raymond. Coordination isomers of biological iron transport compounds III. (1) Transport of λ-cis-chromic deferriferrichrome by Ustilago sphaerogena. Biochem. Biophys. Res. Commun. 60:1066–1071, 1974.

488. Lepp, H., Ed. Geochemistry of Iron. Stroudsburg, Penn.: Dowden, Hutchinson, and Ross, 1975. 464 pp.

489. Lerer, T. J., C. K. Redmond, P. P. Breslin, L. Salvin, and H. W. Rush. Long-term mortality study of steelworkers. VII. Mortality patterns among crane operators. J. Occup. Med. 16:608–614, 1974.

490. Lewis, L. L. Interlaboratory testing programs for the chemical analysis of metals. Standardization News 4(9):19–23, 1976.

491. Lewis, S. M. International Committee for Standardization in Hematology: Proposed recommendations for measurement of serum iron in human blood. Amer. J. Clin. Path. 56:543–545, 1971.

492. Light, P. A., and R. A. Clegg. Metabolism in iron-limited growth, pp. 35–64. In J. B. Neilands, Ed. Microbial Iron Metabolism: A Comprehensive Treatise. New York: Academic Press, 1974.

493. Lintzel, W., J. Rechenberger, and E. Schairer. Über den Eisenstoffwechsel des Neugeborenen und des Säuglings. Z. Gesamte Exp. Med. 113:591–612, 1944.

494. Lipschitz, D. A., T. H. Bothwell, H. C. Seftel, A. A. Wapnick, and R. W. Charlton. The role of ascorbic acid in the metabolism of storage iron. Brit. J. Haematol. 20:155–163, 1971.

495. Lipschitz, D. A., J. D. Cook, and C. A. Finch. A clinical evaluation of serum ferritin as an index of iron stores. New Engl. J. Med. 290:1213–1216, 1974.

496. Lisboa, P. E. Experimental hepatic cirrhosis in dogs caused by chronic massive iron overload. Gut 12:363–368, 1971.

497. Livingston, D. A. Chemical Composition of Rivers and Lakes. U.S. Geological Survey Professional Paper 440-C. Washington, D.C.: U.S. Government Printing Office, 1963. 64 pp.

498. Llinás, M. Metal-polypeptide interactions: the conformational state of iron proteins. Struct. Bond. 17:135–220, 1973.

499. Llinás, M., M. P. Klein, and J. B. Neilands. Solution conformation of the ferrichromes. III. A comparative proton magnetic resonance study of glycine- and serine-containing ferrichromes. J. Mol. Biol. 68:265–284, 1972.

500. Llinás, M., D. M. Wilson, and J. B. Neilands. Effect of metal binding on the conformation of enterobactin. A proton and carbon-13 nuclear magnetic resonance study. Biochemistry 12:3836–3843, 1973.

501. Lloyd, J. W., and A. Ciocco. Long-term mortality study of steelworkers. I. Methodology. J. Occup. Med. 11:299–310, 1969.

502. Lloyd, J. W., F. E. Lundin, Jr., C. K. Redmond, and P. B. Geiser. Long-term mortality study of steelworkers. IV. Mortality by work area J. Occup. Med. 12:151–157, 1970.

503. Lochhead, A. G. Two new species of arthrobacter requiring respectively vitamin B_{12} and the terregens factor. Arch. Mikrobiol. 31:163–170, 1958.

504. Lock, L. F. Iron Deficiency in Plants: How to Control it in Yards and Gardens. Home and Garden Bulletin No. 102. Washington, D.C.: U.S. Department of Agriculture, 1976. 8 pp.

505. Lovenberg, W., Ed. Iron-Sulfur Proteins: Vol. 1. Biological Properties. New York: Academic Press, 1973. 385 pp.

506. Lowe, C. R., H. Campbell, and T. Khosla. Bronchitis in two integrated steel works. III. Respiratory symptoms and ventilatory capacity related to atmospheric pollution. Brit. J. Ind. Med. 27:121–129, 1970.

507. Lucas, D. H. Choosing chimney heights in the presence of buildings, pp. 47–52. In Proceedings of the International Clean Air Conference, Melbourne University, May, 1972. Parkville, Vic., Australia: Clean Air Society of Australia and New Zealand, 1972.

508. Lucas, R. E., and B. D. Knezek. Climatic and soil conditions promoting micronutrient deficiencies in plants, pp. 265–288. In J. J. Mortvedt, P. M.

Giordano, and W. L. Lindsay, Eds. Micronutrients in Agriculture. Madison, Wis.: Soil Science Society of America, 1972.

509. Luckey, M., R. Wayne, and J. B. Neilands. *In vitro* competition between ferrichrome and phage for the outer membrane T5 receptor complex of *Escherichia coli*. Biochem. Biophys. Res. Commun. 64:687–693, 1975.

510. Lundgren, D. G. Microbiological problems in strip mine areas: Relationship to the metabolism of *Thiobacillus ferrooxidans*. Ohio J. Sci. 75:280–287, 1975.

511. Lundin, P. M. The carcinogenic action of complex iron preparations. Brit. J. Cancer 15:838–847, 1961.

512. Lundvall, O. The effect of phlebotomy therapy in porphyria cutanea tarda: Its relation to the phlebotomy-induced reduction of iron stores. Acta Med. Scand. 189:33–49, 1971.

513. Lundvall, O. The effect of replenishment of iron stores after phlebotomy therapy in porphyria cutanea tarda. Acta Med. Scand. 189:51–63, 1971.

514. Lundvall, O., and A. Weinfeld. Iron stores in alcohol abusers. II. As measured with the desferrioxamine test. Acta Med. Scand. 185:271–277, 1969.

515. Lundvall, O., A. Weinfeld, and P. Lundin. Iron storage in porphyria cutanea tarda. Acta Med. Scand. 188:37–53, 1970.

516. Lundvall, O., A. Weinfeld, and P. Lundin. Iron stores in alcohol abusers. I. Liver iron. Acta Med. Scand. 185:259–269, 1969.

517. Luongo, M. A., and S. S. Bjornson. The liver in ferrous sulfate poisoning; A report of three fatal cases in children and an experimental study. New Engl. J. Med. 251:995–999, 1954.

518. Lynch, S. R., I. Berelowitz, H. C. Seftel, G. B. Miller, P. Krawitz, R. W. Charlton, and T. H. Bothwell. Osteoporosis in Johannesburg Bantu males: Its relationship to siderosis and ascorbic acid deficiency. Amer. J. Clin. Nutr. 20:799–807, 1967.

519. Lynch, S. R., H. C. Seftel, J. D. Torrance, R. W. Charlton, and T. H. Bothwell. Accelerated oxidative catabolism of ascorbic acid in siderotic Bantu. Amer. J. Clin. Nutr. 20:641–647, 1967.

520. Lynch, S. R., H. C. Seftel, A. A. Wapnick, R. W. Charlton, and T. H. Bothwell. Some aspects of calcium metabolism in normal and osteoporotic Bantu subjects with special reference to the effects of iron overload and ascorbic acid depletion. South Afr. J. Med. Sci. 35:45–56, 1970.

521. MacDonald, R. A. Hemochromatosis and Hemosiderosis. Springfield, Illinois: Charles C Thomas, 1964. 374 pp.

522. MacDonald, R. A., R. S. Jones, and G. S. Pechet. Folic acid deficiency and hemochromatosis. Arch. Path. 80:153–160, 1965.

523. MacDonald, R. A., and G. S. Pechet. Experimental hemochromatosis in rats. Amer. J. Path. 46:85–109, 1965.

524. Macdougall, L. G. Red cell metabolism in iron deficiency anemia. III. The relationship between glutathione peroxidase, catalase, serum vitamin E, and susceptibility of iron deficient red cells to oxidative hemolysis. J. Pediatr. 80:775–782, 1972.

525. Macon, W. L., and W. J. Pories. The effect of iron deficiency anemia on wound healing. Surgery 69:792–796, 1971.

526. Magnusson, E. E. O. Iron absorption after antrectomy with gastroduodenostomy. Studies on the absorption from food and from iron salt using a double radioiron isotope technique and whole-body counting. Scand. J. Haematol. (Suppl. 26):1–111, 1976.

527. Mailliard, J. A. Iron deficiency following gastric resection. Amer. J. Gastroenterol. 45:109–113, 1966.

528. Markson, J. L., and J. M. Moore. Autoimmunity in pernicious anaemia and iron-deficiency anaemia: A complement-fixation test using human gastric mucosa. Lancet 2:1240–1243, 1962.

529. Martell, A. E. The chemistry of metal chelates in plant nutrition. Soil Sci. 84:13–26, 1957.

530. Martin, D. F. Coordination chemistry of the oceans. Adv. Chem. Ser. 67:255–269, 1967.

531. Martin, E. J., and R. D. Hill. Mine drainage research program of the Federal Water Pollution Control Administration, pp. 46–63. In Second Symposium on Coal Mine Drainage Research, Mellon Institute, 1968. Monroeville, Penn.: Bituminous Coal Research, Inc., 1968.

532. Martínez-Torres, C., and M. Layrisse. Iron absorption from veal muscle. Amer. J. Clin. Nutr. 24:531–540, 1971.

533. Martínez-Torres, C., and M. Layrisse. Nutritional factors in iron deficiency: Food iron absorption. Clin. Haematol. 2:339–352, 1973.

534. Martínez-Torres, C., I. Leets, M. Renzi, and M. Layrisse. Iron absorption by humans from veal liver. J. Nutr. 104:983–933, 1974.

535. Matrone, G., C. Conley, G. H. Wise, and R. K. Waugh. A study of iron and copper requirements of dairy calves. J. Dairy Sci. 40:1437–1447, 1957.

536. Matrone, G., E. L. Thomason, Jr., and C. R. Bunn. Requirements and utilization of iron by the baby pig. J. Nutr. 72:459–465, 1960.

537. McAllen, P. M., N. F. Coghill, and M. Lubran. The treatment of haemochromatosis: With particular reference to the removal of iron from the body by repeated venesection. Q. J. Med. 26:251–276, 1957.

538. McCall, M. G., G. E. Newman, J. R. P. O'Brien, and L. J. Witts. Studies in iron metabolism. 2. The effects of experimental iron deficiency in the growing rat. Brit. J. Nutr. 16:305–323, 1962.

539. McCance, R. A., and E. M. Widdowson. Absorption and excretion of iron. Lancet 2:680–684, 1937.

540. McCormick, R. A., and G. C. Holzworth. Air pollution climatology, pp. 643–700. In A. C. Stern, Ed. Air Pollution. (3rd ed.) Vol. 1. New York: Academic Press, 1976.

541. McIntosh, M. A., and C. F. Earhart. Effect of iron on the relative abundance of two large polypeptides of the *Escherichia coli* outer membrane. Biochem. Biophys. Res. Commun. 70:315–322, 1976.

542. McMullen, T. B., R. B. Faoro, and G. B. Morgan. Profile of pollutant fractions in nonurban suspended particulate matter. J. Air Pollut. Control Assoc. 20:369–372, 1970.

543. Mena, T., K. Horiuchi, K. Burke, and G. C. Cotzias. Chronic manganese poisoning: Individual susceptibility and absorption of iron. Neurology 19:1000–1006, 1969.

544. Mendel, G. A. Studies on iron absorption: I. The relationships between rate of erythropoiesis, hypoxia and iron absorption. Blood 18:727–736, 1961.

545. Mendel, G. A., R. J. Weiler, and A. Mangalik. Studies on iron absorption. II. The absorption of iron in experimental anemias of diverse etiology. Blood 22:450–458, 1973.

546. Meyer-Bertenrath, J. G., and W. Domschke. I. Mitt.: Reversible Ablösung eisenhaltiger chromoproteide von Rattenleber-Ribosomen. Z. Naturforsch. 25B:61–66, 1970.

547. Mikesell, M. E., G. M. Paulsen, R. Ellis, Jr., and A. J. Casady. Iron utilization by efficient and inefficient sorghum lines. Agron. J. 65:77–80, 1973.

548. Miles, A. A., and P. L. Khimji. Enterobacterial chelators of iron: Their occurrence, detection, and relation to pathogenicity. J. Med. Microbiol. 8:477–490, 1975.

549. Miles, L. E. M., D. A. Lipschitz, C. P. Bieber, and J. D. Cook. Measurement of serum ferritin by a 2-site immunoradiometric assay. Anal. Biochem. 61:209–224, 1974.

550. Miller, D. F. (National Research Council Committee on Feed Composition of the Agricultural Board). Composition of Cereal Grains and Forages. Publication 585. Washington, D.C.: National Academy of Sciences, 1958. 663 pp.

551. Miller, M. W. The Pfizer Handbook of Microbial Metabolites. New York: McGraw-Hill, 1961. 772 pp.

552. Minnich, V., A. Okçuoğlu, Y. Tarcon, A. Arcasoy, S. Cin, O. Yörükoğlu, F. Renda, and B. Demirağ. Pica in Turkey. II. Effect of clay upon iron absorption. Amer. J. Clin. Nutr. 21:78–86, 1968.

553. Mitchell, J. H., and G. Blomqvist. Maximal oxygen uptake. New Engl. J. Med. 284:1018–1022, 1971.

554. Modell, C. B. Transfusional haemochromatosis, pp. 230–240. In H. Kief, Ed. Iron Metabolism and Its Disorders. New York: American Elsevier Publishing Co. Inc., 1975.

555. Modell, C. B., and J. Beck. Long-term desferrioxamine therapy in thalassemia. Ann. N.Y. Acad. Sci. 232:201–210, 1974.

556. Moersch, H. J., and H. M. Conner. Hysterical dysphagia. Arch. Otolaryngol. 4:112–117, 1926.

557. Monsen, E. R., and J. D. Cook. Food iron absorption in human subjects. IV. The effects of calcium and phosphate on the absorption of nonheme iron. Amer. J. Clin. Nutr. 29:1142–1148, 1976.

558. Monsen, E. R., I. N. Kuhn, and C. A. Finch. Iron status of menstruating women. Amer. J. Clin. Nutr. 20:842–849, 1967.

559. Moore, C. V. Iron nutrition and requirements. Ser. Haematol. 6:1–14, 1965.

560. Moore, C. V., and R. Dubach. Observations on the absorption of iron from foods tagged with radioiron. Trans. Assoc. Amer. Phys. 64:245–256, 1951.

561. Moore, D. J. A comparison of the trajectories of rising buoyant plumes with theoretical/empirical models. Atmos. Environ. 8:441–457, 1974.

562. Morgan, E. H. Transferrin and transferrin iron, pp. 29–71. In A. Jacobs and M. Worwood, Eds. Iron in Biochemistry and Medicine. New York: Academic Press, 1974.

563. Morgan, E. H., G. Marsaglia, E. R. Giblett, and C. A. Finch. A method of investigating internal iron exchange utilizing two types of transferrin. J. Lab. Clin. Med. 69:370–381, 1967.

564. Morgan, O. S., P. B. B. Gatenby, D. G. Weir, and J. M. Scott. Studies on an iron-binding component in human gastric juice. Lancet 1:861–863, 1969.

565. Morgan, W. K. C., and H. D. Kerr. Pathologic and physiologic studies of welders' siderosis. Ann. Intern. Med. 58:293–204, 1963.

566. Morgan, W. K. C., and A. Seaton. Hematite pneumoconiosis (silicosiderosis or mixed dust fibrosis), pp. 233–240; hematite, pp. 377–378. In Occupational Lung Diseases. Philadelphia: W. B. Saunders Company, 1975.

567. Morris, E. R., and R. Ellis. Isolation of monoferric phytate from wheat bran and its biological value as an iron source to the rat. J. Nutr. 106:753–760, 1976.

568. Mortvedt, J. J., P. M. Giordano, and W. L. Lindsay, Eds. Micronutrients in Agriculture. Proceedings of a Symposium, 1971. Madison, Wisc.: Soil Science Society of America, Inc., 1972. 666 pp.

569. Mros, B., G. Brüschke, and D. Voigt. Eisenmangel als ätiologischer Faktor der Ozaena? Deutsch. Gesundh. 21:2216–2217, 1966.

570. Muirhead, H., and J. Greer. Three-dimensional Fourier synthesis of human deoxyhaemoglobin at 3.5 Å resolution. Nature 228:516–519, 1970.

571. Murphy, J. R., A. M. Pappenheimer, Jr., and S. T. De Borms. Synthesis of diphtheria *tox*-gene products in *Escherichia coli* extracts. Proc. Nat. Acad. Sci., U.S.A. 71:11–15, 1974.

572. Murphy, L. S., and L. M. Walsh. Correction of micronutrient deficiencies with fertilizers, pp. 347–387. In J. J. Mortvedt, P. M. Giordano, and W. L. Lindsay, Eds. Micronutrients in Agriculture. Madison, Wis.: Soil Science Society of America, Inc., 1972.

573. Murphy, T. P., D. R. S. Lean, and C. Nalewajko. Blue-green algae: Their excretion of iron-selective chelators enables them to dominate other algae. Science 192:900–902, 1976.

574. Murray, M. J., and N. Stein. Does the pancreas influence iron absorption? A critical review of information to date. Gastroenterology 51:694–700, 1966.

575. Murray, M. J., and N. Stein. The effect of achylia gastrica in rats on the absorption of dietary iron. Proc. Soc. Exp. Biol. Med. 133:183–184, 1970.

576. Naiman, J. L., F. A. Oski, L. K. Diamond, G. F. Vawter, and H. Schwachman. The gastrointestinal effects of iron-deficiency anemia. Pediatrics 33:83–99, 1964.

577. Nakamura, F. I., and H. H. Mitchell. The utilization for hemoglobin regeneration of the iron salts used in the enrichment of flour and bread. J. Nutr. 25:39–48, 1943.

578. Nason, A., and W. D. McElroy. Modes of action of the essential mineral elements, pp. 451–536. In F. C. Steward, Ed. Plant Physiology: A Treatise. Vol. 3. Inorganic Nutrition of Plants. New York: Academic Press, 1963.

579. Nath, L., S. K. Sood, and N. C. Nayak. Experimental siderosis and liver injury in the rhesus monkey. J. Path. 106:103–111, 1972.

580. National Academy of Sciences. National Academy of Engineering. Environmental Studies Board. Water Quality Criteria 1972. A Report of the Committee on Water Quality Criteria. Washington, D.C.: U.S. Government Printing Office, 1974. 594 pp.

581. National Institute for Occupational Safety and Health. Method P&CAM 173, pp. 173-1–173-8. In NIOSH Manual of Analytical Methods. Cincinnati: U.S. Department of Health, Education, and Welfare, 1974.

582. National Research Council, Agricultural Board. Composition of Concentrate By-Product Feeding Stuffs. Publ. 449. Washington, D.C.: National Academy of Sciences, 1956. 126 pp.

583. National Research Council, Agricultural Board, and Canadian Department of Agriculture. Atlas of Nutritional Data on United States and Canadian Feeds. Washington, D.C.: National Academy of Sciences, 1971. 722 pp.

584. National Research Council, Committee on Animal Nutrition, Agricultural Board, and Department of Agriculture, Canada, Research Branch, Com-

mittee on Feed Composition. United States-Canadian Tables of Feed Composition. Nutritional Data for United States and Canadian Feeds. Second Revision. Publication 1684. Washington, D.C.: National Academy of Sciences, 1969. 92 pp.

585. Necheles, T. F., S. Chung, R. Sabbah, and D. Whitten. Intensive transfusion therapy in thalassemia major: An eight-year follow-up. Ann. N.Y. Acad. Sci. 232:179–185, 1974.

586. Neilands, J. B. Iron and its role in microbial physiology, pp. 3–34. In J. B. Neilands, Ed. Microbial Iron Metabolism: A Comprehensive Treatise. New York: Academic Press, 1974. 597 pp.

587. Neilands, J. B. Microbial iron transport compounds (siderochromes), pp. 167–202. In G. L. Eichhorn, Ed. Inorganic Biochemistry. Vol. 1. New York: Elsevier Scientific Publishing Company, 1973.

588. Neilands, J. B. Microbial iron transport compounds (siderophores), pp. 5–44. In W. F. Anderson and M. C. Hiller, Eds. Proceedings of a Symposium. Development of Iron Chelators for Clinical Use. DHEW Publ. (NIH) 76-994. Bethesda, Md.: U.S. Department of Health, Education, and Welfare, National Institutes of Health, 1976.

589. Neilands, J. B. Naturally occurring non-porphyrin iron compounds. Struct. Bond. 1:59–108, 1966.

590. Neilands, J. B. Siderophores: Biochemical ecology and mechanism of iron transport in enteric bacteria. Adv. Chem. Ser. 162:3–32, 1977.

591. Neilands, J. B., and C. Ratledge. Microbial iron transport compounds (siderophores). In A. I. Laskin and H. A. Lechevalier, Eds. CRC Handbook of Microbiology, Vol. IV. Cleveland: CRC Press, 1979. (in press)

592. Nettesheim, P., D. A. Creasia, and T. J. Mitchell. Carcinogenic and cocarcinogenic effects of inhaled synthetic smog and ferric oxide particles. J. Nat. Cancer Inst. 55:159–169, 1975.

593. Newcomb, E. H. Fine structure of protein-storing plastids in bean root tips. J. Cell Biol. 33:143–163, 1967.

594. Nicholas, D. J. D., and A. Nason. Mechanism of action of nitrate reductase from *Neurospora*. J. Biol. Chem. 211:183–197, 1954.

595. Nicholls, D. Iron, pp. 979–1051; Cobalt, pp. 1053–1107; and Nickel, pp. 1109–1161. In J. C. Bailar, Jr., H. J. Eméleus, R. Nyholm, and A. F. Trotman-Dickenson, Eds. Comprehensive Inorganic Chemistry. Vol. 3. New York: Pergamon Press, 1973.

596. Nicholls, P., and W. B. Elliott. The cytochromes, pp. 221–277. In A. Jacobs and M. Worwood, Eds. Iron in Biochemistry and Medicine. New York: Academic Press, 1974.

597. Nicholson, W. J. Analysis of amphibole asbestiform fibers in municipal water supplies. Environ. Health Perspect. 9:165–172, 1974.

598. Nilsson, L., and Sölvell, L. Clinical studies on oral contraceptives—a randomized doubleblind, crossover study of 4 different preparations (Anovlar® mite, Lyndiol® mite, Ovulen® and Volidan®). Acta Obstet. Gynecol. Scand. 46(Suppl. 8):1–31, 1967.

599. Nissim, J. A. Experimental siderosis: A study of the distributuion, delayed effects and metabolism of massive amounts of various iron preparations. J. Path. Bacteriol. 66:185–204, 1953.

600. Noyes, W. D., F. Hosain, and C. A. Finch. Incorporation of radioiron into marrow heme. J. Lab. Clin. Med. 64:574–580, 1964.

601. Nussbaumer, T., H. C. Pattner, and A. Rywlin. Hémochromatose juvénile chez trois soeurs et un frère avec consanguinité des parents. Etude anatomoclinique et génétique du syndrome endocrino-hépato-myocardique. J. Genet. Hum. 1(2):53–82, 1952.

602. Oades, J. M. The nature and distribution of iron compounds in soils. Soils Fertiliz. 26:69–80, 1963.

603. Oborn, E. T. A survey of pertinent biochemical literature, pp. 111–190, Paper 1459-F. In Chemistry of Iron in Natural Water. U.S. Geological Survey Water-Supply Paper 1459. Washington, D.C.: U.S. Government Printing Office, 1960.

604. Oborn, E. T., and J. D. Hem. Microbiologic factors in the solution and transport of iron, pp. 213–235, Paper 1459-H. In Chemistry of Iron in Natural Water. U.S. Geological Survey Water-Supply Paper 1459. Washington, D.C.: U.S. Government Printing Office, 1961.

605. O'Brien, R. T. Ascorbic acid enhancement of desferrioxamine-induced urinary iron excretion in thalassemia major. Ann. N.Y. Acad. Sci. 232:221–225, 1974.

606. Oertli, J. C., and L. Jacobson. Some quantitative considerations in iron nutrition of higher plants. Plant Physiol. 35:683–688, 1960.

607. Oliver, R. A. M. Siderosis following transfusions of blood. J. Path. Bacteriol. 77:171–194, 1959.

608. Olsen, C. Iron absorption in different plant species as a function of the pH value of the solution. C. R. Trav. Lab. Carlsberg 31:41–59, 1958.

609. Olsen, S. R. Micronutrient interactions, pp. 243–264. In J. J. Mortvedt, P. M. Giordano, and W. L. Lindsay, Eds. Micronutrients in Agriculture. Madison, Wis.: Soil Science Society of America, 1972.

610. Olsen, S. R., F. S. Watanabe, and C. V. Cole. Effect of sodium bicarbonate on the solubility of phosphorus in calcareous soils. Soil Sci. 89:288–291, 1960.

611. Olsson, K. S. Iron stores in normal men and male blood donors: As measured by desferrioxamine and quantitative phlebotomy. Acta Med. Scand. 192:401–407, 1972.

612. Orfei, E., F. I. Volini, F. Madera-Orsini, O. T. Minick, and G. Kent. Effect of iron loading on the hepatic injury induced by ethionine. Amer. J. Path. 52:547–567, 1968.

613. Pakiser, L. C., and R. Robinson. Composition of the continental crust as estimated from seismic observations, pp. 620–626. In J. S. Steinhart and T. J. Smith, Eds. The Earth Beneath the Continents. A Volume in Honor of Merle A. Tuve. (A. G. U. Monograph 10) Washington, D.C.: American Geophysical Union, 1967.

614. Parer, J. T., W. D. Jones, and J. Metcalfe. A quantitative comparison of oxygen transport in sheep and human subjects. Respir. Physiol. 2:196–206, 1967.

615. Park, C. F., Jr., and R. A. MacDiarmid. Ore Deposits. (3rd ed.) San Francisco: W. H. Freeman and Company, 1975. 529 pp.

616. Parker, R. L. Composition of the Earth's Crust. U.S. Geological Survey Professional Paper 440-D. Washington, D.C.: U.S. Government Printing Office, 1967. 19 pp.

617. Parks, G. A. The isoelectric points of solid oxides, solid hydroxides, and aqueous hydroxo complex systems. Chem. Rev. 65:177–198, 1965.

618. Paterson, D. R. A clinical type of dysphagia. J. Laryngol. 34:289–291, 1919.

619. Payne, S. M., and R. A. Finkelstein. Pathogenesis, and immunology of experimental gonococcal infection: Role of iron in virulence. Infect. Immun. 12:1313–1318, 1975.

620. Peacock, W. C., R. D. Evans, J. W. Irvine, Jr., W. M. Good, A. F. Kip, S. Weiss, and J. G. Gibson. The use of two radioactive isotopes of iron in trace studies of erythrocytes. J. Clin. Invest. 25:504–615, 1946.

621. Pearson, W. N., and M. B. Reich. Studies of ferritin and a new iron-binding protein found in the intestinal mucosa of the rat. J. Nutr. 99:137–140, 1969.

622. Peirson, D. H. La teneur de l'air en métaux lourds. Intertecnic 1973(March):192–194.

623. Peirson, D. H., P. A. Cawse, and R. S. Cambray. Chemical uniformity of airborne particulate matter, and a maritime effect. Nature 251:675–679, 1974.

624. Perkin-Elmer Corporation. Analytical Methods for Atomic Absorption Spectrophotometry. Norwalk, Conn.: Perkin-Elmer Corporation, 1971.

625. Perman, G. Hemochromatosis and red wine. Acta Med. Scand. 182:281–284, 1967.

626. Perutz, M. F., H. Muirhead, J. M. Cox, and L. C. G. Goaman. Three-dimensional Fourier synthesis of horse oxyhaemoglobin at 2.8 Å resolution: The atomic model. Nature 219:131–139, 1968.

627. Peters, T. J., and C. A. Seymour. Acid hydrolase activities and lysosomal integrity in liver biopsies from patients with iron overload. Clin. Sci. Mol. Med. 50:75–78, 1976.

628. Peters, W. J., and R. A. J. Warren. Itoic acid synthesis in *Bacillus subtilis*. J. Bacteriol. 95:360–366, 1968.

629. Pettijohn, F. J. Chemical Composition of Sandstones, Excluding Carbonate and Volcanic Sands. U.S. Geological Survey Professional Paper 440-S. Washington, D.C.: U.S. Government Printing Office, 1963. 19 pp.

630. Pettijohn, F. J. Sedimentary Rocks. (3rd ed.) New York: Harper and Row, 1975. 628 pp.

631. Pirzio-Biroli, G., T. H. Bothwell, and C. A. Finch. Iron absorption. II. The absorption of radioiron administered with a standard meal in man. J. Lab. Clin. Med. 51:37–48, 1958.

632. Pirzio-Biroli, G., and C. A. Finch. Iron absorption. III. The influence of iron stores on iron absorption in the normal subject. J. Lab. Clin. Med. 55:216–220, 1960.

633. Plunkett, E. R. Handbook of Industrial Toxicology. New York: Chemical Publishing Co., Inc. 1976. 552 pp.

634. Poldervaart, A. Chemistry of the earth's crust, pp. 119–144. In A. Poldervaart, Ed. Crust of the Earth (A symposium). Geological Society of America Special Paper 62. Baltimore: Waverly Press, 1955.

635. Pollack, J. R., B. N. Ames, and J. B. Neilands. Iron transport in *Salmonella typhimurium*: Mutant blocked in the biosynthesis of enterobactin. J. Bacteriol. 104:635–639, 1970.

636. Pollack, S., S. P. Balcerzak, and W. H. Crosby. Fe[59] absorption in human subjects using total-body counting technic. Blood 28:94–97, 1966.

637. Pollack, S., J. N. George, R. C. Reba, R. Kaufman, and W. H. Crosby. The absorption of nonferrous metals in iron deficiency. J. Clin. Invest. 44:1470–1473, 1965.

638. Pollack, S., and F. D. Lasky. A new iron-binding protein isolated from intestinal mucosa. J. Lab. Clin. Med. 87:670–679, 1976.

639. Pollitt, E., and R. L. Leibel. Iron deficiency and behavior. J. Pediatr. 88:372–381, 1976.

640. Pollycove, M. Iron metabolism and kinetics. Semin. Hematol. 3:235–298, 1966.

641. Pond, W. G., and E. F. Walker, Jr. Cadmium-induced anemia in growing rats: Prevention by oral or parenteral iron. Nutr. Rep. Int. 5:365–370, 1972.

642. Pootrakul, P., A. Christensen, B. Josephson, and C. A. Finch. The role of transferrin in determining internal iron distribution. Blood 49:957–966, 1977.

643. Port, C. D., M. C. Henry, D. G. Kaufman, C. C. Harris, and K. V. Ketels. Acute changes in the surface morphology of hamster tracheobronchial epithelium following benzo[a]pyrene and ferric oxide administration. Cancer Res. 33:2498–2506, 1973.

644. Powell, L. W. Changing concepts in haemochromatosis. Postgrad. Med. J. 46:200–209, 1970.

645. Powell, L. W. Normal human iron storage and its relation to ethanol consumption. Australas. Ann. Med. 15:110–115, 1966.

646. Powell, L. W., R. Mortimer, and O. D. Harris. Cirrhosis of the liver: A comparative study of the four major aetiological groups. Med. J. Austral. 1:941–950, 1971.

647. Pozza, G., and A. Ghidoni. Studies on the diabetic syndrome of idiopathic haemochromatosis. Diabetologia 4:83–86, 1968.

648. Price, C. A. Iron compounds and plant nutrition. Ann. Rev. Plant Physiol. 19:239–248, 1968.

649. Price, C. A., H. E. Clark, and E. A. Funkhouser. Functions of micronutrients in plants, pp. 231–239. In J. J. Mortvedt, P. M. Giordano, and W. L. Lindsay, Eds. Micronutrients in Agriculture. Proceedings of a Symposium, 1971. Madison, Wisc.: Soil Science Society of America, Inc., 1972.

650. Pritchard, J. Significance of iron deficiencies. In reproductive performance, pp. 81–85. In National Research Council, Food and Nutrition Board. Extent and Meanings of Iron Deficiency in the U.S. Proceedings of Workshop, March 8–9, 1971. Washington, D.C.: National Academy of Sciences, 1971.

651. Pritchard, J. A., and R. A. Mason. Iron stores of normal adults and replenishment with oral iron therapy. J.A.M.A. 190:897–901, 1964.

652. Prockop, D. J. Role of iron in the synthesis of collagen in connective tissue. Fed. Proc. 30:984–990, 1971.

653. Puro, D. G., and G. W. Richter. Ferritin synthesis by free and membrane-bound (poly)ribosomes of rat liver. Proc. Soc. Exp. Med. Biol. 138:399–403, 1971.

654. Ragan, H. A., S. Nacht, G. R. Lee, C. R. Bishop, and G. E. Cartwright. Effect of ceruloplasmin on plasma iron in copper-deficient swine. Amer. J. Physiol. 217:1320–1323, 1969.

655. Ramsay, C. A., I. A. Magnus, A. Turnbull, and H. Baker. The treatment of porphyria cutanea tarda by venesection. Q. J. Med. 43:1–24, 1974.

656. Ramsay, W. N. M., and E. A. Campbell. Iron metabolism in the laying hen. Biochem. J. 58:313–317, 1954.

657. Rath, C. E., and C. A. Finch. Sternal marrow hemosiderin: A method for the determination of available iron stores in man. J. Lab. Clin. Med. 33:81–86, 1948.

658. Raymond, K. N. Kinetically inert complexes of the siderophores in studies of microbial iron transport. Adv. Chem. Ser. 162:33–54, 1977.

659. Redmond, C. K., J. Gustin, and E. Kamon. Long-term mortality experience of steelworkers. VIII. Mortality patterns of open hearth steelworkers (A preliminary report). J. Occup. Med. 17:40–43, 1975.

660. Reinhold, J. G., F. Ismail-Beigi, and B. Faradji. Fibre vs. phytate as determinant of the availability of calcium, zinc and iron of breadstuffs, Nutr. Rep. Int. 12:75–85, 1975.

661. Reissmann, K. R., and T. J. Coleman. Acute intestinal iron intoxication. II. Metabolic, respiratory and circulatory effects of absorbed iron salts. Blood 10:46–51, 1955.

662. Reissmann, K. R., T. J. Coleman, B. S. Budai, and L. R. Moriarty. Acute intestinal iron intoxication. I. Iron absorption, serum iron and autopsy findings. Blood 10:35–45, 1955.

663. Reno, H. T. Iron and steel, pp. 667–697. In U.S. Bureau of Mines. Minerals Yearbook 1974. Vol. 1. Metals, Minerals, and Fuels. Washingington, D.C.: U.S. Government Printing Office, 1976.

664. Rhoads, W. A., and A. Wallace. Possible involvement of dark fixation of CO_2 in lime-induced chlorosis. Soil Sci. 89:248–256, 1960.

665. Richmond, V. S., M. Worwood, and A. Jacobs. The iron content of intestinal epithelial cells and its subcellular distribution: Studies on normal, iron-overloaded, and iron-deficient rats. Brit. J. Haematol. 23:605–614, 1972.

666. Rios, E., D. A. Lipschitz, J. D. Cook, and N. J. Smith. Relationship of maternal and infant iron stores as assessed by determination of plasma ferritin. Pediatrics 55:694, 1975.

667. Risdon, R. A., M. Barry, and D. M. Flynn. Transfusional iron overload: The relationship between tissue iron concentration and hepatic fibrosis in the thalassaemia. J. Path. 116:83–95, 1975.

668. Robbins, E., and T. Pederson. Iron: Its intracellular localization and possible role in cell division. Proc. Nat. Acad. Sci. U.S.A. 66:1244–1251, 1970.

669. Roberts, R. J., S. Nayfield, R. Soper, and T. H. Kent. Acute iron intoxication with intestinal infarction managed in part by small bowel resection. Clin. Toxicol. 8:3–12, 1975.

670. Roche, M., and M. Layrisse. The nature and causes of "hookworm anemia." Amer. J. Trop. Med. Hyg. 15:1029–1100, 1966.

671. Rogers, C. H., and J. W. Shive. Factors affecting the distribution of iron in plants. Plant Physiol. 7:227–252, 1932.

672. Ronov, A. B., and A. A. Yaroshevsky. Earth's crust geochemistry, pp. 243–254. In R. W. Fairbridge, Ed. Encyclopedia of Geochemistry and Environmental Sciences. New York: Van Nostrand Reinhold Co., 1969.

673. Rösler, H. J., and H. Lange. Geochemical Tables. (Rev. ed.) New York: Elsevier Publishing Company, 1972. 468 pp. (Translated from the German by H. Liebscher.)

674. Ross, F. G. M. Pyloric stenosis and fibrous stricture of the stomach due to ferrous sulfate poisoning. Brit. Med. J. 2:1200–1202, 1953.

675. Russell, C. S., and W. J. Vaughan. Steel Production: Processes, Products, and Residuals. Baltimore: Johns Hopkins University Press, 1976. 328 pp.

676. Rybo, G. Menstrual blood loss in relation to parity and menstrual pattern. Acta Obstet. Gynecol. Scand. 45(Suppl. 7):1–45, 1966.

677. Rybo, G. Menstrual loss of iron, pp. 163–171. In L. Hallberg, H.-G. Harwerth, and A. Vannotti, Eds. Iron Deficiency: Pathogenesis, Clinical Aspects, Therapy. London: Academic Press, 1970.

678. Saccomanno, G., V. E. Archer, R. P. Saunders, L. A. James, and P. A. Beckler. Lung cancer of uranium miners on the Colorado plateau. Health Phys. 10:1195–1201, 1964.

679. Saffiotti, U., F. Cefis, and L. H. Kob. A method for the experimental induction of bronchogenic carcinoma. Cancer Res. 28:104–113, 1968.

680. Saffiotti, U., R. Montesano, A. R. Sellakumar, and D. G. Kaufman. Respiratory tract carcinogenesis induced in hamsters by different dose

levels of benzo[*a*]pyrene and ferric oxide. J. Nat. Cancer Inst. 49:1199–1204, 1972.

681. Sagone, A. L., and S. P. Balcerzak. Activity of iron-containing enzymes in erythrocytes and granulocytes in thalassemia and iron deficiency. Amer. J. Med. Sci. 259:350–357, 1970.

682. Samoilova, L. M., and V. I. Kireev. Morphological changes in the lungs of animals exposed to the effect of aerosols developing in the course of welding and cast-iron surfacing. Gig. Tr. Prof. Zabol. 19(4):53–54, 1975. (in Russian)

683. Sandell, E. B. Manganese. F. Bio materials, pp. 437–440. In Colorimetric Determination of Traces of Metals. (2nd ed.) New York: Interscience Publishers, Inc., 1950.

684. Sander, O. A. Further observations on lung changes in electric arc welders. J. Ind. Hyg. Toxicol. 26:79–85, 1974.

685. Sanyal, S. K., W. Johnson, B. Jayalakshmamma, and A. A. Green. Fatal "iron-heart" in an adolescent: Biochemical and ultrastructural aspects of the heart. Pediatrics 55:336–341, 1975.

686. Sayers, M. H., S. R. Lynch, R. W. Charlton, T. H. Bothwell, R. B. Walker, and F. Mayet. Iron absorption from rice meals cooked with fortified salt containing ferrous sulfate and ascorbic acid. Brit. J. Nutr. 31:367–375, 1974.

687. Sayers, M. H., S. R. Lynch, R. W. Charlton, T. H. Bothwell, R. B. Walker, and F. Mayet. The effects of ascorbic acid supplementation on the absorption of iron in maize, wheat and soya. Brit. J. Haematol. 24:209–218, 1973.

688. Schade, S. G., B. F. Felsher, G. M. Bernier, and M. E. Conrad. Interrelationship of cobalt and iron absorption. J. Lab. Clin. Med. 75:435–441, 1970.

689. Schade, S. G., B. F. Felsher, B. E. Glader, and M. E. Conrad. Effect of cobalt upon iron absorption. Proc. Soc. Exp. Biol. Med. 134:741–743, 1970.

690. Scheinberg, I. H. The genetics of hemochromatosis. Arch. Intern. Med. 132:126–128, 1973.

691. Schiffer, L. M., D. C. Price, and E. P. Cronkite. Iron absorption and anemia. J. Lab. Clin. Med. 65:316–321, 1965.

692. Schroeder, H. A. The Trace Elements and Man. Some Positive and Negative Aspects. Old Greenwich, Conn.: The Devin-Adair Company, 1973. 180 pp.

693. Schueneman, J. J., M. D. High, and W. E. Bye. Air Pollution Aspects of the Iron and Steel Industry. Public Health Service Publication No. 999-AP-1. Cincinnati, Ohio: U.S. Department of Health, Education, and Welfare, 1963. 129 pp.

694. Schüler, P. Iron alloys, compounds, pp. 740–741. In Encyclopaedia of Occupational Health and Safety. Vol. 1. Geneva: International Labour Office, 1971.

695. Schumacher, H. R., Jr. Hemochromatosis and arthritis. Arth. Rheum. 7:41–50, 1964.

696. Schumacher, H. R., Jr. Ultrastructural characteristics of the synovial membrane in idiopathic haemochromatosis. Ann. Rheum. Dis. 31:465–473, 1972.

697. Schwartz, S. O., and S. A. Blumenthal. Exogenous hemochromatosis resulting from blood transfusions. Blood 3:617–640, 1948.

698. Schwertmann, U. Die Bildung von Eisenoxidmineralen. Fortschr. Mineral. 46:274–285, 1969.

699. Sebesta, W. Ferrous metallurgical processes, pp. 143–169. In A. C. Stern, Ed. Air Pollution (2nd ed.). Vol. 3. New York: Academic Press, 1968.

700. Seckbach, J. Iron content and ferritin in leaves of iron treated *Xanthium pensylvanicum* plants. Plant Physiol. 44:816–820, 1969.

701. Seftel, H. C., K. J. Keeley, C. Isaacson, and T. H. Bothwell. Siderosis in the Bantu: The clinical incidence of hemochromatosis in diabetic subjects. J. Lab. Clin. Med. 58:837–844, 1961.

702. Seftel, H. C., C. Malkin, A. Schmaman, C. Abrahams, S. R. Lynch, R. W. Charlton, and T. H. Bothwell. Osteoporosis, scurvy, and siderosis in Johannesburg Bantu. Brit. Med. J. 1:642–646, 1966.

703. Sehmel, G. A., and F. D. Lloyd. Resuspension of Plutonium at Rocky Flats. Report BNWL-SA-5085. (CONF-740921-11) Richland, Wash.: Battelle Pacific Northwest Laboratories, 1974. 39 pp.

704. Seip, M., and S. Halvorsen. Erythrocyte production and iron stores in premature infants during the first months of life: The anemia of prematurity—etiology, pathogenesis, iron requirement. Acta Paediatr. Scand. 45:600–617, 1956.

705. Sellakumar, A., and P. Shubik. Carcinogenicity of differnet polycyclic hydrocarbons in the respiratory tract of hamsters. J. Nat. Cancer Inst. 53:1713–1719, 1974.

706. Seshadri, R., J. H. Colebatch, P. Gordon, and H. Ekert. Long-term administration of desferrioxamine in thalassaemia major. Arch. Dis. Child. 49:621–626, 1974.

707. Shacklette, H. T., J. C. Hamilton, J. G. Boerngen, and J. M. Bowles. Elemental Composition of Surficial Materials in the Conterminous United States. U.S. Geological Survey Professional Paper 574-D. Washington, D.C.: U.S. Government Printing Office, 1971. 71 pp.

708. Shacklette, H. T., H. I. Sauer, and A. T. Miesch. Geochemical Environments and Cardiovascular Mortality Rates in Georgia. U.S. Geological Survey Professional Paper 574-C. Washington, D.C.: U.S. Government Printing Office, 1970. 39 pp.

709. Shah, B., and B. Belonje. Liver storage iron in Canadians. Amer. J. Clin. Nutr. 29:66–69, 1976.

710. Shahid, M. J., and N. A. Haydar. Absorption of inorganic iron in thalassaemia. Brit. J. Haematol. 13:713–718, 1967.

711. Sharpe, L. M., W. C. Peacock, R. Cooke, and R. S. Harris. The effect of phytate and other food factors on iron absorption. J. Nutr. 41:433–446, 1950.

712. Sheldon, J. H. Haemochromatosis. London: Oxford University Press, 1935. 382 pp.

713. Shum, Y. S., and W. D. Loveland. Atmospheric trace element concentrations associated with agricultural field burning in the Willamette Valley of Oregon. Atmos. Environ. 8:645–655, 1974.

714. Shupe, J. L. Effects of kish on animals, p. 2. In O. S. Cannon, J. L. Shupe, and R. E. Lamborn. The Problems of Airborne Dust and "Kish" in Utah County. Agricultural Experiment Station Research Report 2. Logan: Utah State University, 1972.

715. Siimes, M. A., J. E. Addiego, Jr., and P. R. Dallman. Ferritin in serum: Diagnosis of iron deficiency and iron overload in infants and children. Blood 43:581–590, 1974.

716. Simon, M., P. Franchimont, N. Murie, B. Ferrand, H. van Cauwenberge, and M. Bourel. Study of somatotropic and gonadotropic pituitary function

in idiopathic haemochromatosis (31 cases). Europ. J. Clin. Invest. 2:384–389. 1972.

717. Simpson, F. B., and J. B. Neilands. Siderochromes in Cyanophyceae: Isolation and characteristics of schizokinen from *Anabaena* sp. J. Phycol. 12:44–48, 1976.

718. Sinniah, R. Transfusional siderosis and liver cirrhosis. J. Clin. Path. 22:567–572, 1969.

719. Six, K. M., and R. A. Goyer. The influence of iron deficiency on tissue content and toxicity of ingested lead in the rat. J. Lab. Clin. Med. 79:128–136, 1972.

720. Skinner, C., and C. F. Kenmure. Haemochromatosis presenting as congestive cardiomyopathy and responding to venesection. Brit. Heart J. 35:466–468, 1973.

721. Slade, D. H., Ed. Meteorology and Atomic Energy 1968. TID-24190. Oak Ridge, Tenn.: U.S. Atomic Energy Commission, Division of Technical Information, 1968. 445 pp.

722. Smith, D. W. Steel industry wastes. J. Water Pollut. Control Fed. 48:1287–1293, 1976.

723. Smith, F. C., Jr., J. M. Mathieson, O. Masi, and T. M. Spittler. Simultaneous analysis of dissolved metal pollutants in water. Amer. Lab. 7(12):41–47, 1975.

724. Smith, M. D., and B. Mallett. Iron absorption before and after partial gastrectomy. Clin. Sci. 16:23–34, 1957.

725. Smith, N. J., S. Rosello, M. B. Say, and K. Yeya. Iron stores in the first five years of life. Pediatrics 16:166–173, 1955.

726. Smith, P. F., and A. W. Specht. Heavy-metal nutrition in relation to iron chlorosis of citrus seedlings. Proc. Fla. State Hort. Soc. 65:101–108, 1952.

727. Spedding, D. J. The interaction of gaseous pollutants with materials at the surface of the earth, pp. 213–241. In J. O'M. Bochris, Ed. Environmental Chemistry. New York: Plenum Press, 1977.

728. Spiro, Th. G., and P. Saltman. Polynuclear complexes of iron and their biological implications. Struct. Bond. 6:116–156, 1969.

729. Sprague, G. F. Germ plasm manipulation in the future, pp. 375–395. In J. D. Eastin, F. A. Haskins, C. Y. Sullivan, and C. H. M. Van Bavel, Eds. Physiological Aspects of Crop Yield. Madison, Wisc.: American Society of Agronomy and Crop Science Society of America, 1969.

730. Stamper, J. W., and H. F. Kurtz. Aluminum. Preprint from Bureau of Mines Bulletin 667. A Chapter from Mineral Facts and Problems, 1975 edition. Washington, D.C.: U.S. Government Printing Office, 1975. 30 pp.

731. Stein, M., D. Blayney, T. Feit, T. G. Georgen, S. Micik, and W. L. Nyhan. Acute iron poisoning in children. West. J. Med. 125:289–297, 1976.

732. Steiner, B. A. Ferrous metallurgical operations, pp. 890–929. In A. C. Stern, Ed. Air Pollution. (3rd ed.) Vol. 4. New York: Academic Press, 1977.

733. Stenback, F., A. Ferrero, R. Montesano, and P. Shubik. Synergistic effect of ferric oxide on dimethylnitrosamine carcinogenesis in the Syrian golden hamster. Z. Krebsforsch. 79:31–38, 1973.

734. Stevens, A. R., Jr., D. H. Coleman, and C. A. Finch. Iron metabolism: Clinical evaluation of iron stores. Ann. Intern. Med. 38:199–205, 1953.

735. Stewart, G. R. The regulation of nitrite reductase level in *Lemna minor* L. J. Exp. Bot. 23:171–183, 1972.

736. Stewart, M. J., and J. S. Faulds. The pulmonary fibrosis of haematite miners. J. Path. Bacteriol. 39:233–253, 1934.

737. Stewart, W. B., P. S. Vassar, and R. S. Stone. Iron absorption in dogs during anemia due to acetylphenylhydrazine. J. Clin. Invest. 32:1225–1228, 1953.

738. Stiles, L. W., J. M. Rivers, L. R. Hackler, and D. R. Van Campen. Altered mineral absorption in the rat due to dietary fiber. Fed. Proc. 35:744, 1976. (abstract)

739. Stocks, A. E., and F. I. R. Martin. Pituitary function in haemochromatosis. Amer. J. Med. 45:839–845, 1968.

740. Stocks, A. E., and L. W. Powell. Carbohydrate intolerance in idiopathic haemochromatosis and cirrhosis of the liver. Q. J. Med. 42:733–749, 1973.

741. Stone, W. D. Gastric secretory response to iron therapy. Gut 9:99–105, 1968.

742. Strauss, M. B. Anemia of infancy from maternal iron deficiency in pregnancy. J. Clin. Invest. 12:345–353, 1933.

743. Strom, G. H. Transport and diffusion of stack effluents, pp. 401–501. In A. C. Stern, Ed. Air Pollution. (3rd ed.) Vol. 1. New York: Academic Press, 1976.

744. Stumm, W., and G. F. Lee. Oxygenation of ferrous iron. Ind. Eng. Chem. 53:143–146, 1961.

745. Sturgeon, P. Studies of iron requirements of infants. III. Influence of supplemental iron during normal pregnancy on mother and infant. B. The infant. Brit. J. Haematol. 5:45–55, 1959.

746. Sturgeon, P., and A. Shoden. Total liver storage iron in normal populations of the USA. Amer. J. Clin. Nutr. 24:469–474, 1971.

747. Subramanian, K. N., G. Padmanaban, and P. S. Sarma. Folin-Ciocalteu reagent for the estimation of siderochromes. Anal. Biochem. 12:106–112, 1965.

748. Sulzer, J. L. Significance of iron deficiencies. Effects of iron deficiency on psychological tests in children, pp. 70–76. In National Research Council, Food and Nutrition Board. Extent and Meanings of Iron Deficiency in the U.S. Summary of Proceedings of Workshop, March 8–9, 1971. Washington, D.C.: National Academy of Sciences, 1971.

749. Surgenor, D. M., B. A. Koechlin, and L. E. Strong. Chemical, clinical, and immunological studies on the products of human plasma fractionation: XXXVII. The metal-combining globulin of human plasma. J. Clin. Invest. 28:73–96, 1949.

750. Sussman, M. Iron and infection, pp. 649–679. In A. Jacobs and M. Worwood, Eds. Iron in Biochemistry and Medicine. New York: Academic Press, 1974.

751. Suzman, M. M. Syndrome of anemia, glossitis and dysphagia: Report of eight cases, with special reference to observation at autopsy in one instance. Arch. Intern. Med. 51:1–21, 1933.

752. Sverdrup, H. U., M. W. Johnson, and R. H. Fleming. The Oceans. Their Physics, Chemistry, and General Biology. New York: Prentice-Hall, Inc., 1942. 1087 pp.

753. Swarup, S., S. K. Ghosh, and J. B. Chatterjea. Aconitase activity in iron deficiency. Acta Haematol. 37:53–61, 1967.

754. Symes, A. L., K. Missala, and T. L. Sourkes. Iron- and riboflavin-dependent metabolism of a monoamine in the rat in vivo. Science 174:153–155, 1971.

755. Symes, A. L., T. L. Sourkes, M. B. H. Youdim, G. Gregoriadis, and H. Birnbaum. Decreased monoamine oxidase activity in liver of iron-deficient rats. Can. J. Biochem. 47:999–1002, 1969.

756. Szold, P. D. Plumbism and iron deficiency. New Engl. J. Med. 290:520, 1974. (letter to Editor)

757. Tait, G. H. The identification and biosynthesis of siderochromes formed by *Micrococcus denitrificans*. Biochem. J. 146:191–204, 1975.

758. Taljaard, J. J. F., B. C. Shanley, W. M. Deppe, and S. M. Joubert. Porphyrin metabolism in experimental hepatic siderosis in the rat. II. Combined effect of iron overload and hexachlorobenzene. Brit. J. Haematol. 23:513–519, 1972.

759. Tan, A. T., and R. C. Woodworth. Ultraviolet difference spectral studies of conalbumin complexes with transition metal ions. Biochemistry 8:3711–3716, 1969.

760. Taras, M. J., A. E. Greenberg, R. D. Hoak, and M. C. Rand, Eds. Standard Methods for the Examination of Water and Wastewater. (13th ed.) Washington, D.C.: American Public Health Association, 1971. 874 pp.

761. Teculescu, D., and A. Albu. Pulmonary function in workers inhaling iron oxide dust. Int. Arch. Arbeitsmed. 31:163–170, 1973.

762. Tentative method of analysis for iron content of atmospheric particulate matter (bathophenanthroline method) 12126-01-70T, pp. 299–301; and Tentative method of analysis for iron content of atmospheric particulate matter (volumetric method) 12126-02-70T, pp. 302-304. In Methods of Air Sampling and Analysis. Washington, D.C.: American Public Health Association, 1972.

763. Theis, T. L., and P. C. Singer. Complexation of iron(II) by organic matter and its effect on iron(II) oxygenation. Environ. Sci. Technol. 8:569–573, 1974.

764. Thompson, R. J., G. B. Morgan, and L. J. Purdue. Analysis of selected elements in atmospheric particulate matter by atomic absorption. Atom. Absorpt. Newslett. 9:53–57, 1970.

765. Thomson, A. B. R., D. Olatunbosun, and L. S. Valberg. Interrelation of intestinal transport system for manganese and iron. J. Lab. Clin. Med. 78:642–655, 1971.

766. Thomson, A. B. R., C. Shaver, D. J. Lee, B. L. Jones, and L. S. Valberg. Effect of varying iron stores on site of intestinal absorption of cobalt and iron. Amer. J. Physiol. 220:674–678, 1970.

767. Thomson, A. B. R., and L. S. Valberg. Intestinal uptake of iron, cobalt, and manganese in iron-deficient rat. Amer. J. Physiol. 223:1327–1329, 1972.

768. Thomson, A. B. R., and L. S. Valberg. Kinetics of intestinal iron absorption in the rat: Effect of cobalt. Amer. J. Physiol. 220:1080–1085, 1970.

769. Thomson, A. B. R., L. S. Valberg, and D. G. Sinclair. Competitive nature of the intestinal transport mechanism for cobalt and iron in the rat. J. Clin. Invest. 50:2384–2394, 1971.

770. Thorne, D. W., and H. B. Peterson. Irrigated Soils. Their Fertility and Management. (2nd ed.) New York: The Blakiston Company, Inc., 1954. 392 pp.

771. Thorne, D. W., F. B. Wann, and W. Robinson. Hypotheses concerning lime-induced chlorosis. Soil Sci. Soc. Amer. Proc. 15:254–258, 1950.

772. Tiffin, L. O. Iron translocation. I. Plant culture, exudate sampling, iron-citrate analysis. Plant Physiol. 41:510–514, 1966.

773. Tiffin, L. O. Iron translocation. II. Citrate/iron ratios in plant stem exudates. Plant Physiol. 41:515–518, 1966.

774. Tiffin, L. O. Translocation of iron citrate and phosphorus in xylem exudate of soybean. Plant Physiol. 45:280–283, 1970.

775. Tiffin, L. O. Translocation of micronutrients in plants, pp. 199–229. In J. J. Mortvedt, P. M. Giordano, and W. L. Lindsay, Eds. Micronutrients in Agriculture. Proceedings of a Symposium, 1971. Madison, Wisc.: Soil Science Society of America, Inc., 1972.

776. Tiffin, L. O., and J. C. Brown. Absorption of iron from iron chelate by sunflower roots. Science 130:274–275, 1959.

777. Tisdale, S. L., and W. L. Nelson. Soil Fertility and Fertilizers. (3rd ed.) New York: Macmillan Publishing Company, Inc., 1975. 694 pp.

778. Torrance, J. D., R. W. Charlton, A. Schmaman, S. R. Lynch, and T. H. Bothwell. Storage iron in "muscle." J. Clin. Path. 21:495–500, 1968.

779. Toth, S. J. The physical chemistry of soils, pp. 142–162. In F. E. Bear, Ed. Chemistry of the Soil. (2nd ed.) (American Chemical Society Monograph Series No. 160) New York: Reinhold Publishing Corp., 1964.

780. Trump, B. F., J. M. Valigorsky, A. U. Arstila, W. J. Mergner, and T. D. Kinney. The relationship of intracellular pathways of iron metabolism to cellular iron overload and the iron storage diseases. Amer. J. Path. 72:295–336, 1973.

781. Turnbull, A. Iron absorption, pp. 369–403. In A. Jacobs and M. Worwood, Eds. Iron in Biochemistry and Medicine. New York: Academic Press, 1974.

782. Turnbull, A., H. Baker, B. Vernon-Roberts, and I. A. Magnus. Iron metabolism in porphyria cutanea tarda and in erythropoietic protoporphyria. Q. J. Med. 42:341–355, 1973.

783. Turner, D. B. Workbook of Atmospheric Dispersion Estimates. Office of Air Programs Publ. AP-26. Research Triangle Park, N.C.: U.S. Environmental Protection Agency, 1970. 88 pp.

784. Twyman, E. S. The effect of iron supply on the yield and composition of leaves of tomato plants. Plant Soil 10:375–388, 1959.

785. Ullrey, D. E., E. R. Miller, O. A. Thompson, I. M. Ackermann, D. A. Schmidt, J. A. Hoefer, and R. W. Luecke. The requirement of the baby pig for orally administered iron. J. Nutr. 70:189–192, 1960.

786. Underwood, E. J. The Mineral Nutrition of Livestock. Aberdeen: The Central Press, Ltd., 1966. 237 pp.

787. Underwood, E. J. Trace Elements in Human and Animal Nutrition. (3rd ed.) New York: Academic Press, 1971. 543 pp.

788. United Kingdom Ministry of Housing and Local Government. U.K. Clean Air Act 1956. Memorandum on Chimney Heights. (2nd ed.) London: H. M. Stationery Office, 1967.

789. U.S. Department of Health, Education and Welfare. Public Health Service. Preliminary Findings of the First Health and Nutrition Examination Survey, United States, 1971–1972. Dietary Intake and Biochemical Findings. DHEW Publ. (HRA)74-1219-1. Rockville, Md.: U.S. Department of Health, Education, and Welfare, 1974. 183 pp.

790. U.S. Department of Health, Education, and Welfare. Ten-State Nutrition Survey 1968–1970. IV. Biochemical. DHEW Public. No. (HSM) 72-8132. Atlanta, Georgia: U.S. Department of Health, Education, and Welfare, Center for Disease Control, (1972). 296 pp.

791. U.S. Department of the Interior. Federal Water Pollution Control Administration. The Cost of Clean Water. Vol. 3. Industrial Waste

Profiles No. 1—Blast Furnace and Steel Mills. FWPCA Publ. I.W.P.-1. Washington, D.C.: U.S. Goverment Printing Office, 1967. 100 pp.

792. U.S. Environmental Protection Agency. Air Pollution Aspects of Emission Sources: Iron and Steel Mills—A Bibliography with Abstracts. Publ. AP-107. Research Triangle Park, N.C.: U.S. Environmental Protection Agency, 1972. 84 pp.

793. U.S. Environmental Protection Agency. Compilation of Air Pollutant Emission Factors. (2nd ed.) Publ. AP-42. Research Triangle Park, N.C.: U.S. Environmental Protection Agency, 1973. 292 pp.

794. U.S. Environmental Protection Agency. Methods Development and Quality Assurance Research Laboratory. Iron (STORET NO. Total 01045), pp. 110–111. In Methods for Chemical Analysis of Water and Wastes. EPA-625-/6-74-003. Washington, D.C.: Office of Technology Transfer, U.S. Environmental Protection Agency, 1974.

795. U.S. Environmental Protection Agency. National Atmospheric Data Bank. Data File on Iron. Research Triangle Park, N.C.: Monitoring and Data Analysis Division, Office of Air Quality Planning and Standards, U.S. Environmental Protection Agency. (Data are continuously accumulated, available from EPA.)

796. U.S. Environmental Protection Agency. Office of Technology Transfer. Methods for Chemical Analysis of Water and Wastes. EPA 625/6-74-003. Washington, D.C.: U.S. Environmental Protection Agency, 1974. 298 pp.

797. U.S. Environmental Protection Agency. Progress in the Prevention and Control of Air Pollution in 1975. Annual Report of the Administrator of the Environmental Protection Agency to the Congress of the United States in Compliance with Section 313 of Public Law 91-604, The Clean Air Act, as Amended. Senate Document No. 94-228. Washington, D.C.: U.S. Government Printing Office, 1976. 171 pp.

798. Valberg, L. S., J. Ludwig, and D. Olatunbosun. Alterations in cobalt absorption in patients with disorders of iron metabolism. Gastroenterology 56:241–251, 1969.

799. Valberg, L. S., J. Sorbie, W. E. N. Corbett, and J. Ludwig. Cobalt test for the detection of iron deficiency anemia. Ann. Intern. Med. 77:181–187, 1972.

800. Van Dyke, D. C., H. O. Anger, and Y. Yano. Progress in determining bone marrow distribution in vivo. Prog. Atom. Med. 2:65–84, 1968.

801. Varian Techtron. Iron section. In Analytical Methods for Flame Spectroscopy. Palo Alto, Calif.: Varian Techtron, 1972. 18 pp.

802. Vilter, R. W. Ascorbic acid. IX. Effects of deficiency in human beings, pp. 348–380. In W. H. Sebrell, Jr., and R. S. Harris, Eds. The Vitamins: Chemistry, Physiology, Pathology, (Vol. 1). New York: Academic Press, 1954.

803. Vinogradov, A. P. The Geochemistry of Rare and Dispersed Chemical Elements in Soils. (2nd ed.) New York: Consultants Bureau, Inc., 1959. 209 pp.

804. Vinson, P. P. Hysterical dysphagia. Minn. Med. 5:107–108, 1922.

805. Viteri, F. E., and B. Torún. Anaemia and physical work capacity. Clin. Haematol. 3:609–626, 1974.

806. Waddell, J., H. F. Sassoon, K. D. Fisher, and C. J. Carr. A Review of the Significance of Dietary Iron on Iron Storage Phenomena. Bethesda, Md.: Federation of American Societies for Experimental Biology, Life Sciences Research Office, 1972. 81 pp.

807. Wagman, D. D., W. H. Evans, V. B. Parker, I. Halow, S. M. Bailey, and

R. H. Schumm. Selected Values of Chemical Thermodynamic Properties. Tables for Elements 35 Through 53 in the Standard Order of Arrangement. National Bureau of Standards Technical Note 270-4. Washington, D.C.: U.S. Government Printing Office, 1969. 141 pp.

808. Waid, J. S. Hydroxamic acids in soil systems, pp. 65–101. In E. A. Paul and A. D. McLaren, Eds. Soil Biochemistry. Vol. 4. New York: Marcel Dekker, 1975.

809. Walker, A. R. P., and U. B. Arvidsson. Iron "overload" in the South African Bantu. Trans. Roy. Soc. Trop. Med. Hyg. 47:536–548, 1953.

810. Walker, R. J., I. W. Dymock, I. D. Ansell, E. B. D. Hamilton, and R. Williams. Synovial biopsy in haemochromatosis arthropathy: Histological findings and iron deposition in relation to total body iron overload. Ann. Rheum. Dis. 31:98–102, 1972.

811. Walker, R. J., and R. Williams. Haemochromatosis and iron overload, pp. 589–612. In A. Jacobs and M. Worwood, Eds. Iron in Biochemistry and Medicine. New York: Academic Press, 1974.

812. Wallace, A., and O. R. Lunt. Iron chlorosis in horticultural plants, a review. Amer. Soc. Hort. Sci. 75:819–841, 1960.

813. Wallace, A., R. T. Mueller, O. R. Lunt, R. T. Ashcroft, and L. M. Shannon. Comparisons of five chelating agents in soils, in nutrient solutions, and in plant responses. Soil Sci. 80:101–108, 1955.

814. Wallerstein, R. O., and P. M. Aggeler. Diagnosis of anemia in childhood. J. Chron. Dis. 12:377–390, 1960.

815. Walters, G. O., F. M. Miller, and M. Worwood. Serum ferritin concentration and iron stores in normal subjects. J. Clin. Path. 26:770–772, 1973.

816. Wann, E. V., and W. A. Hills. The genetics of boron and iron transport in the tomato. J. Hered. 64:370–371, 1973.

817. Wann, F. B. Control of Chlorosis in American Grapes. Agricultural Experiment Station Bulletin 299. Logan: Utah State Agricultural College, 1941. 27 pp.

818. Wanta, R. C. Meteorology and air pollution, pp. 187–226. In A. C. Stern, Ed. Air Pollution. (2nd ed.) Vol. 1. New York: Academic Press, 1968.

819. Wapnick, A. A., T. H. Bothwell, and H. Seftel. The relationship between serum iron levels and ascorbic acid stores in siderotic Bantu. Brit. J. Haematol. 19:271–276, 1970.

820. Wapnick, A. A., S. R. Lynch, P. Krawitz, H. C. Seftel, R. W. Charlton, and T. H. Bothwell. Effects of iron overload on ascorbic acid metabolism. Brit. Med. J. 3:704–707, 1968.

821. Wapnick, A. A., S. R. Lynch, H. C. Seftel, R. W. Charlton, T. H. Bothwell, and J. Jowsey. The effect of siderosis and ascorbic acid depletion on bone metabolism, with special reference to osteoporosis in the Bantu. Brit. J. Nutr. 25:367–376, 1971.

822. Warnock, R. E. Micronutrient uptake and mobility within corn plants (Zea mays L.) in relation to phosphorus-induced zinc deficiency. Soil Sci. Soc. Amer. Proc. 34:765–769, 1970.

823. Warren, R. A. J., and J. B. Neilands. Mechanism of microbial catabolism of ferrichrome A. J. Biol. Chem. 240:2055–2058, 1965.

824. Watt, B. K., and A. Merrill. Composition of Foods: Raw, Processed, Prepared. Agriculture Handbook No. 8. Washington, D.C.: U.S. Government Printing Office, 1963. 190 pp.

825. Wawzkiewicz, E. J., H. A. Schneider, B. Starcher, J. Pollack, and J. B. Neilands. Salmonellosis pacifarin activity of enterobactin. Proc. Nat. Acad. Sci. U.S.A. 68:2870–2873, 1971.

826. Wayne, R., K. Frick, and J. B. Neilands. Siderophore protection against colicins M, B, V, and Ia in *Escherichia coli*. J. Bacteriol. 126:7–12, 1976.

827. Wayne, R., and J. B. Neilands. Evidence for common binding sites for ferrichrome compounds and bacteriophage φ80 in the cell envelope of *Escherichia coli*. J. Bacteriol. 121:497–503, 1975.

828. Webb, T. E., and F. A. Oski. Iron deficiency anemia and scholastic achievement in young adolescents. J. Pediatr. 82:827–830, 1973.

829. Weinberg, E. D. Iron and susceptibility to infectious disease. Science 184:952–956, 1974.

830. Weinfeld, A. Iron stores, pp. 329–372. In L. Hallberg, H. G. Harwerth, and A. Vannotti, Eds. Iron Deficiency: Pathogenesis, Clinical Aspects, Therapy. New York: Academic Press, 1970.

831. Weinfeld, A. Storage iron in man. Acta Med. Scand. 177(Suppl. 427):1–155, 1964.

832. Weinstein, L. H., and W. R. Robbins. The effect of different iron and manganese nutrient levels on the catalase and cytochrome oxidase activities of green and albino sunflower leaf tissues. Plant Physiol. 30:27–32, 1955.

833. Weintraub, L. R., M. E. Conrad, and W. H. Crosby. Regulation of the intestinal absorption of iron by the rate of erythropoiesis. Brit. J. Haematol. 11:432–438, 1965.

834. Weintraub, L. R., M. E. Conrad, and W. H. Crosby. The significance of iron turnover in the control of iron absorption. Blood 24:19–24, 1964.

835. Weiss, M. G. Inheritance and physiology of efficiency in iron utilization in soybeans. Genetics 28:253–268, 1943.

836. Welch, R. M., and D. R. Van Campen. Iron availability to rats from soybeans. J. Nutr. 105:253–256, 1975.

837. Westberg, K., N. Cohen, and K. W. Wilson. Carbon monoxide: Its role in photochemical smog. Science 171:1013–1015, 1971.

838. Westlin, W. F. Deferoxamine in the treatment of acute iron poisoning: Clinical experiences with 172 children. Clin. Pediatr. 5:531–535, 1966.

839. Wetzel, R. G. Limnology. Philadelphia: W. B. Saunders Company, 1975. 743 pp.

840. Wheby, M. S., G. E. Suttle, and K. T. Ford. Intestinal absorption of hemoglobin iron. Gastroenterology 58:647–656, 1970.

841. Wheby, M. S., and G. Umpierre. Effect of transferrin saturation on iron absorption in man. New Engl. J. Med. 271:1391–1395, 1964.

842. Whelpdale, D. M., and R. E. Munn. Global sources, sinks, and transport of air pollution, pp. 289–324. In A. C. Stern, Ed. Air Pollution. (3rd ed.) Vol. 1. New York: Academic Press, 1976.

843. Whitehead, J. S. W., and R. M. Bannerman. Absorption of iron by gastrectomized rats. Gut 5:38–44, 1965.

844. Whitson, T. C., and E. E. Peacock, Jr. Effect of alpha alpha dipyridyl on collagen synthesis on healing wounds. Surg. Gynecol. Obstet. 128:1061–1064, 1969.

845. Whittemore, D. O., and D. Langmuir. The solubility of ferric oxyhydroxides in natural waters. Ground Water 13:360–365, 1975.

846. Whitten, C. F., and A. J. Brough. The pathophysiology of acute iron poisoning. Clin. Toxicol. 4:585–595, 1971.

847. Whitten, C. F., G. W. Gibson, M. H. Good, J. F. Goodwin, and A. J. Brough. Studies in acute iron poisoning. 1. Desferrioxamine in the treatment of acute iron poisoning: clinical observations, experimental studies, and theoretical considerations. Pediatrics 36:322–335, 1965.

848. Widdowson, E. M., and R. A. McCance. Iron exchanges of adults on white and brown bread diets. Lancet 1:588–591, 1942.

849. Widdowson, E. M., R. A. McCance, and C. M. Spray. The chemical composition of the human body. Clin. Sci. 10:113–125, 1951.

850. Williams, R., F. Manenti, H. S. Williams, and C. S. Pitcher. Iron absorption in idiopathic haemochromatosis before, during, and after venesection therapy. Brit. Med. J. 2:78–81, 1966.

851. Williams, R., P. J. Scheuer, and S. Sherlock. The inheritance of idiopathic haemochromatosis: A clinical and liver biopsy of 16 families. Q. J. Med. 31:249–265, 1962.

852. Williams, R., P. M. Smith, E. J. F. Spicer, M. Barry, and S. Sherlock. Venesection therapy in idiopathic haemochromatosis: An analysis of 40 treated and 18 untreated patients. Q. J. Med. 38:1–16, 1969.

853. Williams, R., H. S. Williams, P. J. Scheuer, C. S. Pitcher, E. Loiseau, and S. Sherlock. Iron absorption and siderosis in chronic liver disease. Q. J. Med. 36:151–166, 1967.

854. Williams, R. E. Waste Production and Disposal in Mining, Milling, and Metallurgical Industries. San Francisco: Miller Freeman Publications, Inc., 1975. 489 pp.

855. Wintrobe, M. M. Clinical Hematology. (7th ed.) Philadelphia: Lea and Febiger, 1974. 1896 pp.

856. Witts, L. J. Simple achlorhydric anaemia. Guy's Hosp. Rep. 80:253–296, 1930.

857. Witzleben, C. L., and B. E. Buck. Iron overload hepatotoxicity: A postulated pathogenesis. Clin. Toxicol. 4:579–583, 1971.

858. Witzleben, C. L., and N. J. Chaffey. Response of the iron-loaded liver to prolonged ethionine feeding. Arch. Path. 80:447–452, 1965.

859. Witzleben, C. L., and J. P. Wyatt. The effect of long survival on the pathology of thalassaemia major. J. Path. Bacteriol. 82:1–12, 1961.

860. Woodruff, C. W., and E. B. Bridgeforth. Relationship between the hemogram of the infant and that of the mother during pregnancy. Pediatrics 12:681–685, 1953.

861. Woodworth, R. C., K. G. Morallee, and R. J. P. Williams. Perturbations of the proton magnetic resonance spectra of conalbumin and siderophilin as a result of binding Ga^{3+} or Fe^{3+}. Biochemistry 9:839–842, 1970.

862. World Health Organization. Nutritional Anemias. Report of a WHO Group of Experts. WHO Technical Report Series No. 503. Geneva: World Health Organization, 1972. 29 pp.

863. World Health Organization. Nutritional Anemias. Report of a WHO Scientific Group. WHO Technical Report Series No. 405. Geneva: World Health Organization, 1968. 37 pp.

864. Worwood, M., A. Edwards, and A. Jacobs, Non-ferritin iron compound in rat small intestinal mucosa during iron absorption. Nature 229:409–410, 1971.

865. Wretlind, A. Food iron supply, pp. 39–64. In L. Hallberg, H.-G. Harwerth, and A. Vannotti, Eds. Iron Deficiency: Pathogenesis, Clinical Aspects, Therapy. New York: Academic Press, 1970.

866. Wright, G. W. The Effects of Inhaled Iron Oxide on the Health of Man. Paper 62–73 Presented at the 55th Annual Meeting, Air Pollution Control Association, Chicago, Ill., May, 1962. 8 pp.

867. Wutscher, H. K., E. O. Olson, A. V. Shull, and A. Peynado. Leaf nutrient levels, chlorosis, and growth of young grapefruit trees on 16 rootstocks grown on calcareous soil. J. Amer. Soc. Hort. Sci. 95:259–261, 1970.

868. Wynder, E. L., S. Hultberg, F. Jacobsson, and I. J. Bross. Environmental factors in cancer of the upper alimentary tract: A Swedish study with special reference to Plummer-Vinson (Paterson-Kelly) syndrome. Cancer 10:470–487, 1957.

869. Yancy, R. J., Jr., S. A. L. Breeding, and C. E. Lankford. Enterochelin: A Virulence Factor for *Salmonella typhimurium*. Paper B 82 Presented at American Society for Microbiology, 76th Annual Meeting, Atlantic City, N.J., May 2–7, 1976.

870. Youdim, M. B. H., H. F. Woods, B. Mitchell, D. G. Grahame-Smith, and S. Callender. Human platelet monoamine oxidase activity in iron-deficiency anaemia. Clin. Sci. Mol. Med. 48:289–295, 1975.

871. Zadeh, J. A., C. D. Karabus, and J. Fielding. Haemoglobin concentration and other values in women using an intrauterine device or taking corticosteroid contraceptive pills. Brit. Med. J. 4:708–711, 1967.

872. Zagainov, V. A., A. G. Sutugin, I. V. Petryanov-Solokov, and A. A. Lushnikov. Probability of adhesion of molecular aerosol particles to surfaces. Dokl. Akad. Nauk SSSR 221:367–369, 1975. (in Russian)

873. Zajic, J. E. Microbial Biogeochemistry. New York: Academic Press, 1969. 345 pp.

Index